Praise for *Disappearing Through the Skylight*

"A smooth and fascinating read . . . Hardison shows that intelligence, invention, and humanism still thrive."

—*Chicago Tribune*

"Engaging and provocative . . . *Disappearing Through the Skylight* is quite a performance."

—*The Philadelphia Inquirer*

"An intelligent antidote to the age of apocalypse"

—*Kirkus Reviews*

"It is a pleasure bordering on ecstasy to have a book on science and technology and its potential for creative achievement written by an author with the impeccable humanistic and scholarly credentials of O. B. Hardison."

—*Houston Chronicle*

"Exhilarating reading . . . Hardison wraps his first-class mind around science, architecture, poetry, art, and computers."

—*Newsday*

PENGUIN BOOKS

DISAPPEARING THROUGH THE SKYLIGHT

O. B. Hardison, Jr., was born in 1928 and educated at the University of North Carolina at Chapel Hill and at the University of Wisconsin. For many years the director of the Folger Shakespeare Library in Washington, D.C., Hardison was also an influential cultural critic and scholar. His works include *Entering the Maze: Identity and Change in Modern Culture*, *Toward Freedom and Dignity: The Humanities and the Idea of Humanity*, and *Prosody and Purpose in the English High Renaissance*, as well as two collections of poetry. The recipient of scholastic and literary honors from more than twenty universities and societies, he was awarded the Italian Government's Cavaliere Ufficiale in 1974 and the Order of the British Empire in 1983. Hardison was also a founding member of the Quark Club, a group of scientists and humanists interested in cultural change. In 1980 he was John F. Kennedy visiting lecturer to New Zealand and, most recently, University Professor at Georgetown University in Washington, D.C., where he lived with his wife, Marifrancis Fitzgibbon, until his death in August of 1990.

DISAPPEARING

CULTURE AND

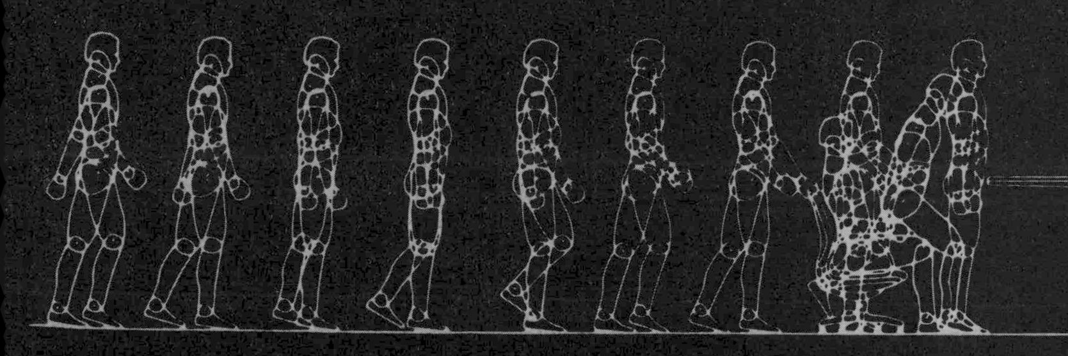

THROUGH THE SKYLIGHT

TECHNOLOGY IN THE TWENTIETH CENTURY

O. B. HARDISON, Jr.

PENGUIN BOOKS

PENGUIN BOOKS
Published by the Penguin Group
Viking Penguin, a division of Penguin Books USA Inc.,
375 Hudson Street, New York, New York 10014, U.S.A.
Penguin Books Ltd, 27 Wrights Lane,
London W8 5TZ, England
Penguin Books Australia Ltd, Ringwood,
Victoria, Australia
Penguin Books Canada Ltd, 2801 John Street,
Markham, Ontario, Canada L3R 1B4
Penguin Books (N.Z.) Ltd, 182–190 Wairau Road,
Auckland 10, New Zealand

Penguin Books Ltd, Registered Offices:
Harmondsworth, Middlesex, England

First published in the United States of America by
Viking Penguin, a division of Penguin Books USA Inc., 1989
Published in Penguin Books 1990

10 9 8 7 6 5 4 3 2 1

Portions of this book first appeared in somewhat
different form in *The Georgia Review*, *The Pushcart Prize XII*, *The Sewanee Review*, and the published
proceedings of the 1984 conference
"Science and Literature."

Pages 391-392 constitute an extension of
this copyright page.

LIBRARY OF CONGRESS CATALOGING IN PUBLICATION DATA
Hardison, O. B.
Disappearing through the skylight: culture and technology in the
twentieth century/O. B. Hardison, Jr.
p. cm.
Includes bibliographical references (p.) and index.
ISBN 0 14 01.1582 X
1. Technology and civilization. 2. Arts—History—20th century.
3. Human evolution. 4. Nature. 5. Culture I. Title.
[CB478.H36 1990]
303.48′3′0904—dc20 90–7552

Printed in the United States of America

FOR JAMES FITZGIBBON, A.I.A.

Architect, Teacher, and Friend

1915–1985

I am captivated more by dreams of the future
than by the history of the past.

—THOMAS JEFFERSON

PREFACE

Disappearing Through the Skylight is about change in modern culture. It examines five basic and interrelated areas—nature, history, language, art, and human evolution—reviewing the ways in which central concepts in each area have changed since the beginning of the present century. Because the changes have been fundamental, the concepts—and even the vocabularies and images in which the concepts tend to be framed—no longer seem to objectify a real world. It is as though progress were making the real world invisible.

No examination of modern culture can exclude the influence of science and technology, and one that understated their influence would be irresponsible. I have devoted much attention to both in the pages that follow, and, if only for that reason, I should state at the outset that this is not a technical book. My point of view is that of an interested and involved citizen of modern culture, not that of a scientist or historian of science or technology. I have tried to inform myself on major developments in areas I have treated. I also confess to being fascinated and dazzled by the brilliance of the achievements of modern science, and I hope my admiration is evident in what I have written. This book is not, however, an attempt to explain or popularize this or that development in science or technology. It is concerned with values rather than processes or products, and is intended for all those readers, nonscientists and scientists alike, who are interested in understanding the culture we inhabit.

My method is simple. I have tried to listen carefully to what people involved in creating and interpreting cultural change say about what they are doing. The testimony of artists has been as important in this respect as the testimony of scientists. Art is always in some sense a comment on experience, and a defining theme of twentieth-century art, whether "applied," as in the case of architecture, or "fine," as in the case of painting and poetry, is response to radical change. Artists have often welcomed the new ideas and techniques of this century as means of liberation from a dying—and deadening—past; they have also, and probably almost as often, responded to change with nostalgia, anguish, and despair.

Change is normally thought of as something that happens in the outside world. In 1920, for example, there were no television sets. By 1980, television sets were out there in the world wherever you looked. However, change is always subjective as well as objective. The mind is always shaped by its experience of the world. A person born into a world without television sees things differently from a person born into a world in which television is commonplace.

There is another fact about cultural innovations. If an innovation is basic, simply because it *is* so, a generation after it has been introduced, it becomes part of the world as given—part of the shape of consciousness, you might say, rather than the content of consciousness. Television seemed to the generation of the 1940s to be an amazing triumph of human ingenuity and pregnant with social implications. Its influence on modern culture has been every bit as great as those who witnessed its birth imagined. But now that it has been assimilated, it is taken for granted, which means that, in a sense, it has become obvious and invisible at the same time. Anyone who discusses modern culture has to do a great deal of contemplating of the invisible in the obvious.

Modern culture is often said to be fragmented. The growth of knowledge produces branchings and subbranchings of specialties in the same way that evolution produces branchings and subbranchings of species. As the branchings multiply, they seem to isolate people in procedures and languages that are unintelligible to people in other specialties. The most familiar version of this critique is C. P. Snow's argument that progress has split modern culture in half so that there are two cultures—one scientific and the other nonscientific—and that there is little communication between them.

The idea of cultural fragmentation has its uses, but it is misleading as a guide to the way people experience culture. Consciousness is never a series of separate departments. It is a unity, a single identity that is the product of all of the influences that have shaped it, which is to say that it is the end result of everything that makes up the culture. To use an analogy, when you have eaten a piece of cake, you do not say, "The flour and sugar and eggs and vanilla taste good." You say, "The cake tastes good." By the same token, there are not fifteen cultures or even two cultures; there is one culture. It is the culture we inhabit. It shaped our consciousness as we grew up, and it continues to influence us throughout life in ways we perceive and in ways of which we are largely unconscious.

Taken together, the five areas I have singled out for treatment provide something like a cross section of modern culture. They are usually thought of as separate, and each is normally treated in isolation from the others. Obviously, however, they are interrelated, and new developments in one reverberate in all the others. Darwin explains the theory of natural selection in a prose that is laden with aesthetic and ethical values and is frequently poetic. A few years later, Ferdinand Brunetière, a critic, writes a history of literature entitled *The Evolution of Literary Genres*. Having created a new branch of mathematics called fractal geometry, the twentieth-century mathematician Benoit Mandelbrot draws conclusions from it about aesthetics and produces images so daring and beautiful that they insist on being considered aesthetically as well as scientifically. More generally, the theory of nature offered by physics is obviously of central importance to an artist who wishes to represent nature—or "imitate" it, as Leonardo might have said—in poetry or painting.

Nature, history, language, and art are parts of a wonderfully intricate mobile: touch one and the rest tremble and change position in sympathy. I make no claim to having invented this idea, but one of the delights of writing *Disappearing Through the Skylight* was rediscovering how pervasive, surprising, and subtle the interrelations are.

To make my text more readable I decided not to use footnotes. I have, however, always tried to make the source of important quotations and citations clear in context, and I have included a chapter-by-chapter bibliography at the end. Using this, the reader should have no difficulty tracking down material that is of special interest and finding suggestions for further reading.

This book continues an analysis begun with *Entering the Maze: Identity and Change in Modern Culture* (1982). Its ideas were first sketched out in 1978 in a series of lectures on cultural change delivered at Clemson University at the invitation of a committee of that university chaired by Dr. Richard Calhoun. Since then, I have continued to extend and, I hope, deepen my thinking, and, of course, important developments have continued to occur in all of the areas treated.

A preliminary version of the discussion of language—"The Poetry of Nothing"—was delivered at the conference "Science and Literature," convened by Dr. John Broderick of the Library of Congress in 1984, and was published in the "proceedings" volume of that conference in 1985. Other parts of this book have appeared in somewhat different form in *The Georgia Review* and *The Sewanee Review*, and I am grateful to their editors, Mr. Stanley Lindberg and Mr. George Core, respectively, for permission to use them. I also thank Mr. Bill Henderson, editor of the Pushcart Press, for publication of an early version of the discussion of Darwin, D'Arcy Thompson, and Benoit Mandelbrot in Pushcart Prize volume XII.

For my understanding of computer art I am heavily indebted to Paul Trachtman, a senior editor of the *Smithsonian Magazine*, and my exposure to cybernetics and machine intelligence owes much to meetings between 1981 and 1984 of an informal group of scientists and humanists organized by Paul calling itself "The Quark Club."

One member of the club, Dr. Paul MacLean of the National Institutes of Health and the author of *A Triune Concept of the Brain and Behavior* (1973), introduced me to contemporary speculation about the evolutionary basis of mental processes and about parallels between organic and machine evolution. My understanding of current research on the brain was deepened by conversations with my colleague at Georgetown University, Dr. Richard M. Restak, author of *The Brain: An Exploration of the Human Mind and Our Future* (1979).

Three decades of immensely stimulating conversations with James Fitzgibbon, late of the School of Architecture of Washington University in St. Louis, helped me gain a perspective on the course of architecture in the twentieth century. The dedication of this book to him will suggest how deep my obligations are, although, of course, I accept full responsibility for all facts and opinions here offered.

A period during which I was privileged to work closely with Warren Cox of the Washington, D.C., architectural firm of Hartman and Cox, helped me understand where and how postmodernist architecture fits into the larger picture. My appreciation of the visionary aspect of design and of the relationship between design and the efficient use of materials was enriched by a friendship of more than two decades with the late R. Buckminster Fuller.

For many details about the history of modern painting, I owe an obvious debt to the magnificent collections of the National Gallery of Art, the Hirshhorn Museum, and the National Air and Space Museum, all part of the Smithsonian Institution in Washington, D.C. My wife, Marifrancis, taught me many years ago to appreciate the intricacies of the style of Gertrude Stein.

Illustrations have always been an integral part of my plan of this book. It is a pleasure to express my gratitude to Lesa Hartjens of Imagefinders in Washington, D.C., for invaluable assistance in locating and securing permissions for the illustrations in the present volume. I owe a still larger debt to Amanda Vaill of Viking Penguin for her expert editorial work.

WASHINGTON, D.C.
AUGUST 27, 1988

CONTENTS

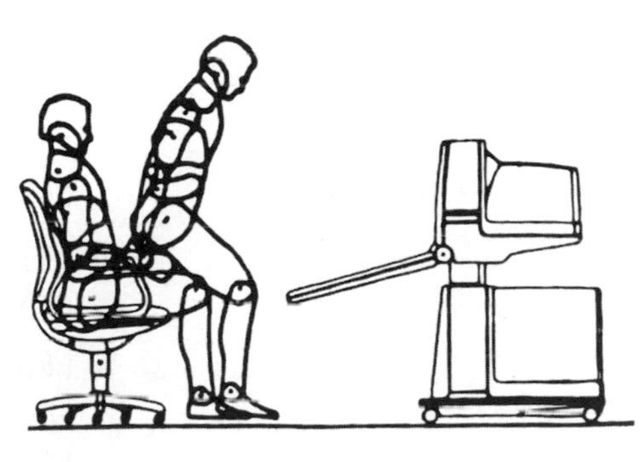

Color illustrations appear between pages 136 and 137.

DISAPPEARING THROUGH THE SKYLIGHT

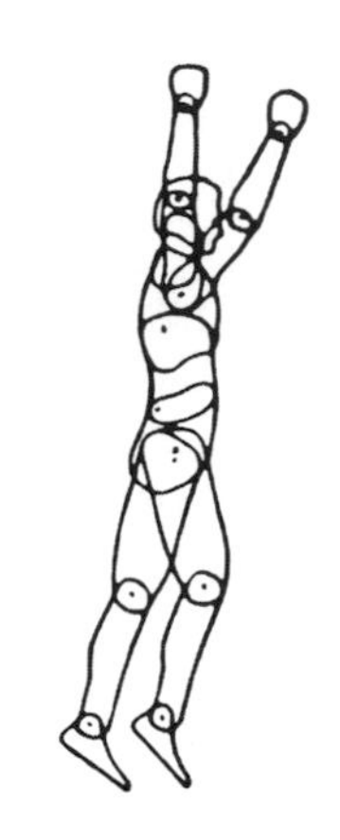

PROLOGUE

In the past, people not mistakenly assumed that the world would always be the same. Today this prognosis is manifestly false; but we are still in the dark about what our world will become.
—CARL FRIEDRICH VON WEIZSÄCKER,
The Unity of Nature

In the nineteenth century, science presented nature as a group of objects set comfortably and solidly in the middle distance before the eyes of the beholder. In the work of D'Arcy Thompson, published around the turn of the century, nature has disappeared. It has become a set of geometric and mathematical relations that lie under the surface of the visible. It is still, however, indubitably there. Today, nature has slipped, perhaps finally, beyond our field of vision. We can imitate it in mathematics—we can even produce convincing images of it—but we can never know it. We can know only our own creations.

This book is about the ways culture has changed in the past century, changing the identities of all of those born into it. Its metaphor for the effect of change on culture is "disappearance."

Because of its close alliance with technology, architecture objectifies the forms of modern culture with great clarity. Since the turn of the century, a global architectural style reflecting the state of the art at the moment of design has begun to replace local and traditional styles. Thus a suspension bridge has the same form whether over a gorge in the Himalayas or the mouth of San Francisco Bay. A Hilton Hotel offers essentially the same accommodations whether in Tokyo or Denver, and the visible similarities are objective manifestations of invisible similarities in, for example, administrative structures and accounting and financial policies. By the same token, a McDonald's hamburger is the same in New York and Rome, and a Pepsi-Cola produces the same bubbles in Vladivostok and Grand Rapids. In all of these cases, the effect of change has been the disappearance of regional and parochial identities and the emergence of a global consciousness.

The universalizing process of technology has touched every facet of culture. As this has happened, the sense of individuality has inevitably faded.

Visual art sensed the new, universalist thrust of culture in the first decade of the twentieth century. We can see the human image caught in the act of disappearing in drawings and paintings by Picasso from 1901, when he presented Señora Canals in an orthodox realistic style, to 1906, when he began to experiment with distortions and generalizations of the human image based on primitive Iberian and African art, to 1911, when, in paintings like *Ma Jolie*, the human image has disappeared entirely.

The movement reached its first climax in the elegant geometrical abstractions of Piet Mondrian and the geometric formalism of Bauhaus architecture and the so-called International Style. Other art forms emphasized aspects of the same universalizing aesthetic. The highly finished metal sculptures of Jean Arp, Constantin Brancusi, and José de Rivera objectify in nonutilitarian form the finely shaped mathematical surfaces of propellers and turbine blades and airfoils. Christo's *Running Fence* objectifies the aesthetic of continuous structures, an aesthetic of framing and reticulation that is implicit in the railroad tracks, high tension lines, and highways that segment and contain the modern landscape.

In spite of the success of the visual arts in expressing an abstract aesthetic, language seems to resist the trend. It seems to lag behind the

Pablo Picasso, portrait of Señora Canals (1901)—early representational style.

other arts, alternating between baffled silence and the celebration of exhausted pieties.

The problem of language came powerfully to the surface of literary art in the last poem of Stéphane Mallarmé, "Un Coup de dés" ("A Throw of the Dice"), published at the turn of the century. The poem is about overcoming silence. What is most striking about it is the power of the silence against which it struggles. To overcome the silence, Mallarmé must, in effect, invent a new language that uses words in nonverbal ways—as visual counters—and treats the page as part of the process of expression rather than as a neutral ground for blocks of type. His new language thus overcomes silence by abandoning the old forms of language—by assimilating characteristics of arts different from the art of language as traditionally practiced. Mallarmé's title is also significant. Randomness—a throw of the dice—is a central concern of twentieth-century art and parallels a similar concern of twentieth-century science. Perhaps incorporating randomness is a way to reestablish a living relationship between language and the world. Perhaps . . . and perhaps not. Too much randomness seems to make language dig in its heels. Randomness causes disruptions that push language away from what most people consider "meaning." The result is something like an equivalent in language to abstraction in art, but does such a concept make sense? Can language exist without meaning, and if it could, what good would it be?

Efforts to create an art of language equivalent to abstraction in visual art include Dada, sound poems, concrete poetry, and the work of Oulipo—*Ouvroir de la littérature potentielle*—which displaces grammar by mathematical formulas and combinatory procedures to create new forms of potential expression. None of these experiments has aroused much enthusiasm in the literary establishment. Even though many of the strategies explored by Dada and concrete poetry have been used to good effect in commercial art and magazine and television advertising, and have demonstrated by their success that they communicate to the general public, their significance has been ignored. Language is still too deeply associated with meaning—hence with the sense of the past and stability of identity—to be as flexible as visual imagery. To admit that meaning can be separated from language seems to surrender an anchor holding man to the earth.

Meanwhile, a new art form that works with images, sounds, and text, separately and in combination, has appeared. Computer art cannot avoid the perspective of technology because it is the child of technology.

Computer art is holistic in its simultaneous use of image, sound, and text, and it is often kinetic. It moves and changes. It reaches out to surround and absorb the consumer, creating an artificial reality that forces the consumer to confront the increasing irrelevance in modern culture of the distinction between the real, in the sense of that which occurs nat-

Pablo Picasso, portrait of Gertrude Stein (1906)—simplified and formalized style.

urally, and the artificial, in the sense of that which is a human artifact.

In computer art, the consumer often controls the product. Sometimes this is by choice, as in a computer novel where the reader decides whether to enter the cave or not. Sometimes the control is involuntary, as when the brain waves of the listener are returned through feedback to the machine and determine the shape of the music produced. In such cases, the distinction between producer and consumer, composer and listener—or between author and reader—is blurred. The listener becomes the composer and the reader the author. Computer art also marks a fundamental change in the concept of art as property. A computer artwork need not be "owned" like earlier art. In principle, it can be given away with no more loss than is involved in "giving" someone the telephone number of the nearest EXXON gas station.

Computers now share the human environment. Most obviously they exhibit rudimentary intelligence. They also have been equipped with arms and grippers and legs, and in this form they have begun to act physically on the world around them and modify it. Inevitably, they affect the sense of human identity. Is the mind a machine—and a relatively simple one at that, once the trick of programming with neurons is understood? Is the claim of humanity to uniqueness disappearing along with the claim of each human to a separate identity shaped by a local habitation and a name? Is the idea of what it is to be human disappearing, along with so many other ideas, through the modern skylight?

In its fearless exploration of inner and outer worlds, modern culture has evidently reached a turning point—a kind of phase transition from one set of values to another. Crossing the barrier that separates the phases is another kind of disappearance.

The nature of that barrier is nicely characterized in a phrase developed by science in connection with the search for extraterrestrial life: "horizon of invisibility."

A horizon of invisibility cuts across the geography of modern culture. Those who have passed through it cannot put their experience into familiar words and images because the languages they have inherited are inadequate to the new worlds they inhabit. They therefore express themselves in metaphors, paradoxes, contradictions, and abstractions rather than languages that "mean" in the traditional way—in assertions that are apparently incoherent or collages using fragments of the old to create

Pablo Picasso, Ma Jolie *(1911)—in which the subject has all but disappeared.*

enigmatic symbols of the new. The most obvious case in point is modern physics, which confronts so many paradoxes that physicists like Paul Dirac and Werner Heisenberg have concluded that traditional languages are, for better or worse, simply unable to represent the world that science has forced on them. In "Quantum Mechanics and a Talk with Einstein," Heisenberg remarks, "I assume that the mathematical scheme works, but no link with the traditional language has been established so far." The same comment might be made about the relation between the twentieth-century languages of Cubism, collage, Dada, and concrete poetry and the visual and verbal languages that preceded them.

Many people have already passed beyond the barrier separating the phases of modern culture. They are different—odd—perhaps like the converts of the fourth century A.D. who crossed the barrier between pagan and Christian culture. The only way these converts could express their experience was through paradoxes and impossibilities. They spoke of being "reborn." They claimed to have experienced "a peace that passes all understanding." They claimed that we see here through a glass darkly and only later will be able to see truth face-to-face. In festive moods they announced that the wisdom of God is the foolishness of man; but they refused absolutely to make simple, direct sense. Perhaps that was because with the best will in the world, they couldn't make sense. They were speaking a language that was accurate for their experience but out of phase with the language of those who had not crossed the barrier with them.

It was "like" speaking a new and therefore unintelligible language or rising from the dead or being a child again. The unconverted considered these figures of speech absurd, and that is precisely the point. In his *Confessions* Saint Augustine labeled the culture he had left a "land of unlikeness" because whatever had been taken over in the new culture had been transformed. There was nothing "like" the old in the new culture, and by the same token, there was nothing "like" the new culture in the old. The normal strategy of explaining something that is strange is to compare it to something familiar that it is like. Since nothing on one side of Saint Augustine's barrier was "like" anything on the other, there was no way of communicating between the two worlds. The simple truths of each language sounded like nonsense to speakers of the other. In this sense, those who had crossed the barrier were truly invisible to those who had not.

The transition that is occurring in modern culture is something like this process, although it obviously does not involve religious conversion. What it *does* involve is movement into the unknown. The experience is sometimes frightening and often confusing, but it can also be exciting and challenging. It is, at any rate, dynamic and associated with life, not death. The fourth century was a time of anguish for those who regretted the loss of the past and of hope for those who looked to the future. If the fourth century is any precedent for what is happening in modern culture, the present transition will be immensely creative as well as painful and will involve a great unfolding as well as a transformation of the human spirit.

At no other time in history have conditions been more exciting—or more filled with promise—for beings on a small planet hurtling through the vast darkness of space toward an unknown and unimaginable future which they, themselves, are creating.

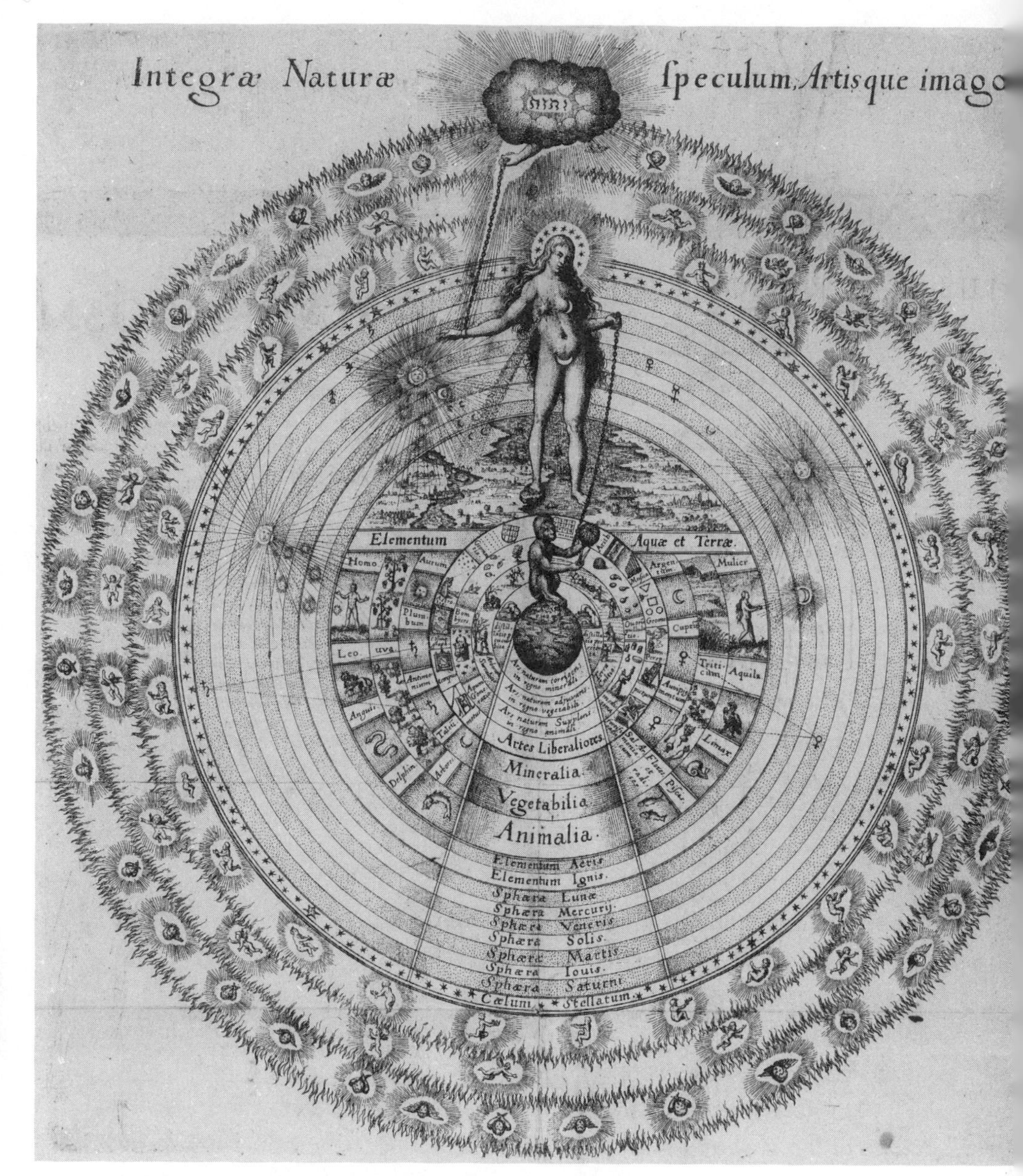

Robert Fludd, Ptolemaic Image of Nature *(1624), showing the earth at the center surrounded by the spheres of the planets and fixed stars, and beyond them, the invisible world of the angels.*

PART I

A TREE, A STREAMLINED FISH, AND A SELF-SQUARED DRAGON; OR, THE DISAPPEARANCE OF NATURE

From harmony, from heav'nly harmony,
This universal frame began.
When Nature underneath a heap
Of jarring atoms lay,
And could not heave her head,
The tuneful voice was heard from high,
"Arise, ye more than dead."
Then cold and hot and moist and dry
In order to their stations leap,
And Music's power obey.
From harmony, from heav'nly harmony
This universal frame began;
From harmony to harmony
Through all the compass of the notes it ran,
The diapason closing full in man. . . .

—JOHN DRYDEN,
"Ode for St. Cecilia's Day"

1

THE MUSIC OF THE WORLD

The root meaning of the Greek *kosmos*—from which are derived English "cosmos" and also "cosmetic"—is "order" or "arrangement." It has the additional meanings of "comely order," "decoration," or "ornament."

In its dominant mode, Greek science was the study of the comely and harmonious order of the world—an order that is in the world whether it is perceived or not. Science, for the Greeks, was more an aesthetic than a practical pursuit. Its great triumphs were in geometry and the theory of proportions.

Greek science assumed that creation is beautiful as well as orderly. The proper response to it is aesthetic as well as intellectual. In the *Theogony* Hesiod (seventh century B.C.) imagines that creation is pervaded by the music of the nine muses, the daughters of Mnemosyne: "Unwearying flows the sweet sound from their lips, and the house of their father Zeus the loud-thunderer is glad at the lilylike voice of the goddesses as it spreads abroad, and the peaks of snowy Olympus resound, and the homes of the immortals." The world is music, and creation a dance.

For more than two thousand years, this orderly and comely beauty was symbolized by the nine concentric spheres determined by the orbits of the fixed stars, the planets, the sun, and the moon and revolving perpetually around the earth. The invisible world begins at the outer surface of the last sphere. Christian tradition peopled it with nine orders of angels. In the rendering of this idea (p. 8) by the seventeenth-century philosopher and mystic Robert Fludd, the goddess Nature stands on the earth, her head, which is crowned with stars, extending through the sphere dividing the invisible from the visible world, and touches the angels. The legs of a compass descend from the center of the earth. Their points touch the inside of the sphere that encloses the visible world. The compass symbolizes the ancient faith that the secrets of nature are rational and reveal themselves to man through science.

In the sixth century B.C. Pythagoras of Samos discovered the relation between harmony and mathematical ratios—according to legend, after hearing the tones produced by blacksmith's hammers of different weights. His discovery convinced him that number is the foundation of the comely order of the world and that harmony is its corollary. The natural order is simultaneously functional and beautiful, and its basis is number. Aristotle explains that the Pythagoreans "supposed . . . numbers to be the elements of all things, and the whole of heaven to be a musical scale and a number." The Pythagorean scale accounts for the songs of the angels and the music of the spheres and also for the diapason that extends from God to man, creating the music of the world so elegantly illustrated in Robert Fludd's *Musica mundana*.

The idea that the world is a comely and beautiful order remained central in later Greek scientific thought. It may, in fact, have eventually hampered the development of Greek science by encouraging interest in elegant demonstration at the expense of calculation. The motto over the doors of Plato's academy was *Medeis ageometretos eisito*—"Let no one ignorant of geometry enter." In Book VII of *The Republic*, Socrates argues

Robert Fludd,
Musica mundana
(1624).

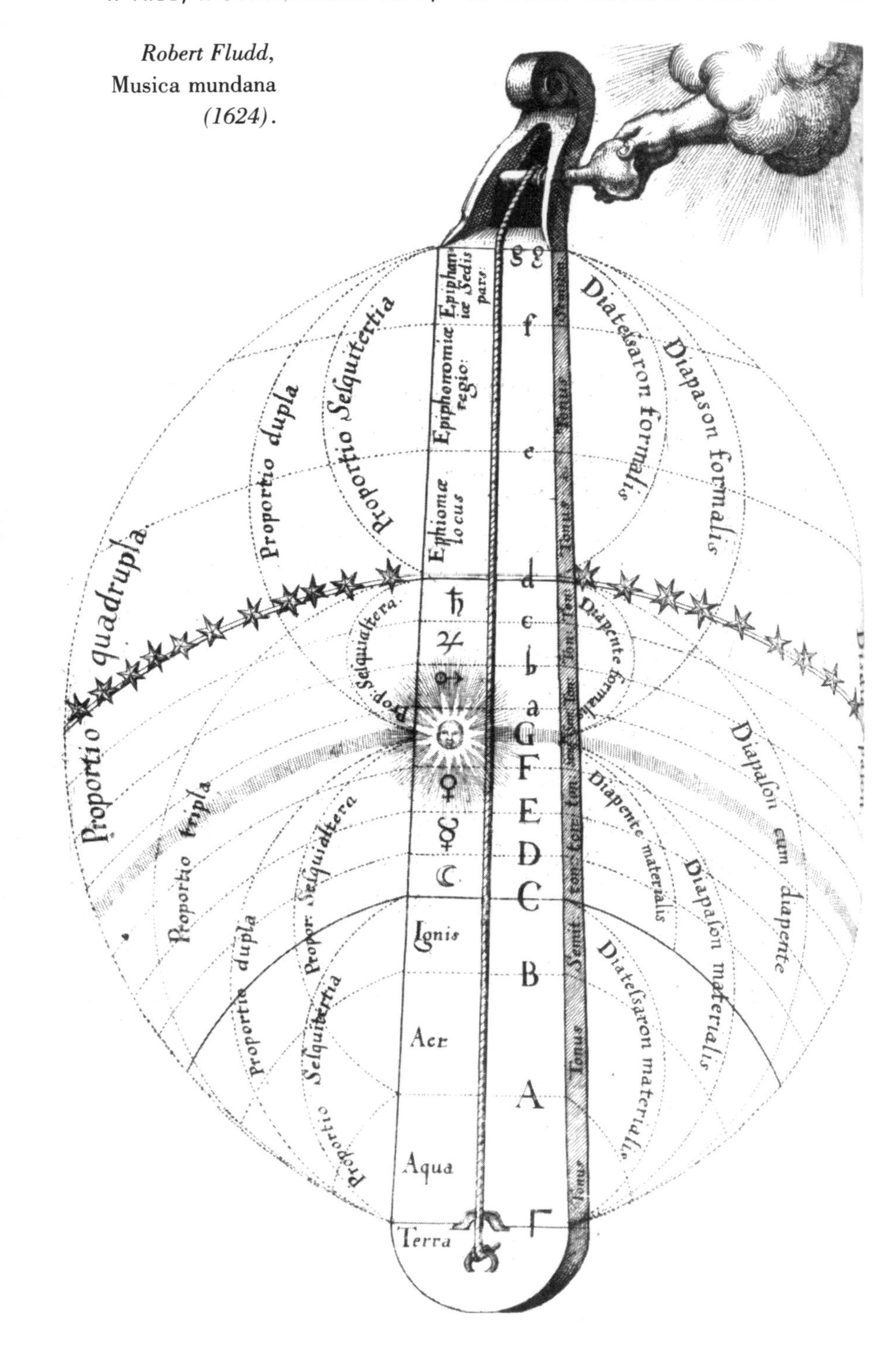

that arithmetic and geometry, the sciences of number and figure, are the basis of rational understanding of the real:

> These sparks that paint the sky, since they are decorations on a visible surface, we must consider the first and most exact of visible things, but we must recognize that they fall far short of the truth—the movements, namely—of real slowness and real speed in true numbers and true figures. . . . These can be apprehended by reason and thought, not by sight.

This insistence on an essential, unseen, but discoverable order has persisted in later scientific thought. Christian apologists could (and did) discover Pythagorean notes in the Old Testament as well as in Plato. Of wisdom Proverbs says: "When he prepared the heavens, I was there: when he set a compass upon the face of the depth." In Book VII of Milton's *Paradise Lost* (first ed., 1667), this becomes:

> . . . in his hand
> He took the golden Compasses, prepar'd
> In God's Eternal store, to circumscribe
> This Universe, and all created things:
> One foot he centered, and the other turn'd
> Round through the vast profundity obscure,
> Thus far extend, thus far thy bounds,
> This be thy just Circumference, O world.

The Pythagorean note rings again in the work of the great Renaissance astronomer Johannes Kepler (1571–1630). In the *Timaeus* (53–56) Plato explains how the four elements, which seem to be irreducible, are in reality expressions of mathematical form. The smallest particles—earth—are cubes; the element of water is composed of almost-spherical twenty-sided particles (icosahedra) whose roundness accounts for liquidity. Air is made up of eight-sided figures (octahedra), and fire is made up of pyramids (tetrahedra). Yet even the elements are not the ground of the real. The elements are themselves made up of triangles, and the more complex elements are created from simpler combinations.

Early in his career Kepler concluded that the same sort of geometry that governs the four elements governs the structure of the solar system.

He decided specifically that the distances between the planets can be determined by assuming they are the distances that would be produced by inscribing the five regular solids, beginning with the tetrahedron, inside of each other. Thus if a cube is inscribed within the orbit of Saturn, the orbit of Jupiter will just fit within it, and so forth. Evidently, God has so planned things that the smallest particles, the alpha of creation, are also involved in its completion, its omega:

> It is my intention, Reader, to demonstrate that the Highest and Most Good Creator in the creation of this mobile world and the arrangement of the heavens had his eye on those five regular bodies [i.e., solids] which have been celebrated from the time of Pythagoras and Plato down to our own day; and that to their nature He accommodated the number of the heavenly spheres, their proportions, and the system of their motions.

The visual result of Kepler's analysis is a remarkable picture of the cosmos that was included in his *Cosmographic Mystery* (1596; see p. 16).

By coincidence, the best measurements of the distances of the planetary orbits available to Kepler, those made by the astronomer Tycho Brahe, corresponded closely with the conclusions reached by Kepler through his geometric analysis. He thus felt his conclusions were empirically confirmed. It is interesting to speculate what might have happened if the observed measurements had not been so close to the ones he reached deductively. Which would Kepler have abandoned—his Pythagorean geometry or Tycho Brahe's painstaking measurements?

The Pythagorean scientist is drawn to the world as much by love of its endlessly surprising and beautiful patterns as by the desire for knowledge. He is a connoisseur as well as an observer of facts. If he does not create beauty, he discovers it and shares it with others. At the end of his *Harmony of the World* (1619), Kepler cries, "I thank thee, Lord God and Creator, that you have permitted me to see the beauty of your work of creation." Michael Faraday, a nineteenth-century bookbinder turned scientist, discovered electromagnetic lines of force and used the discovery to invent the dynamo. In a twentieth-century discussion of Faraday's experiments, J. B. S. Haldane touches the Pythagorean chord: "As a result of Faraday's work you are able to listen to a wireless. But more than that, as a result

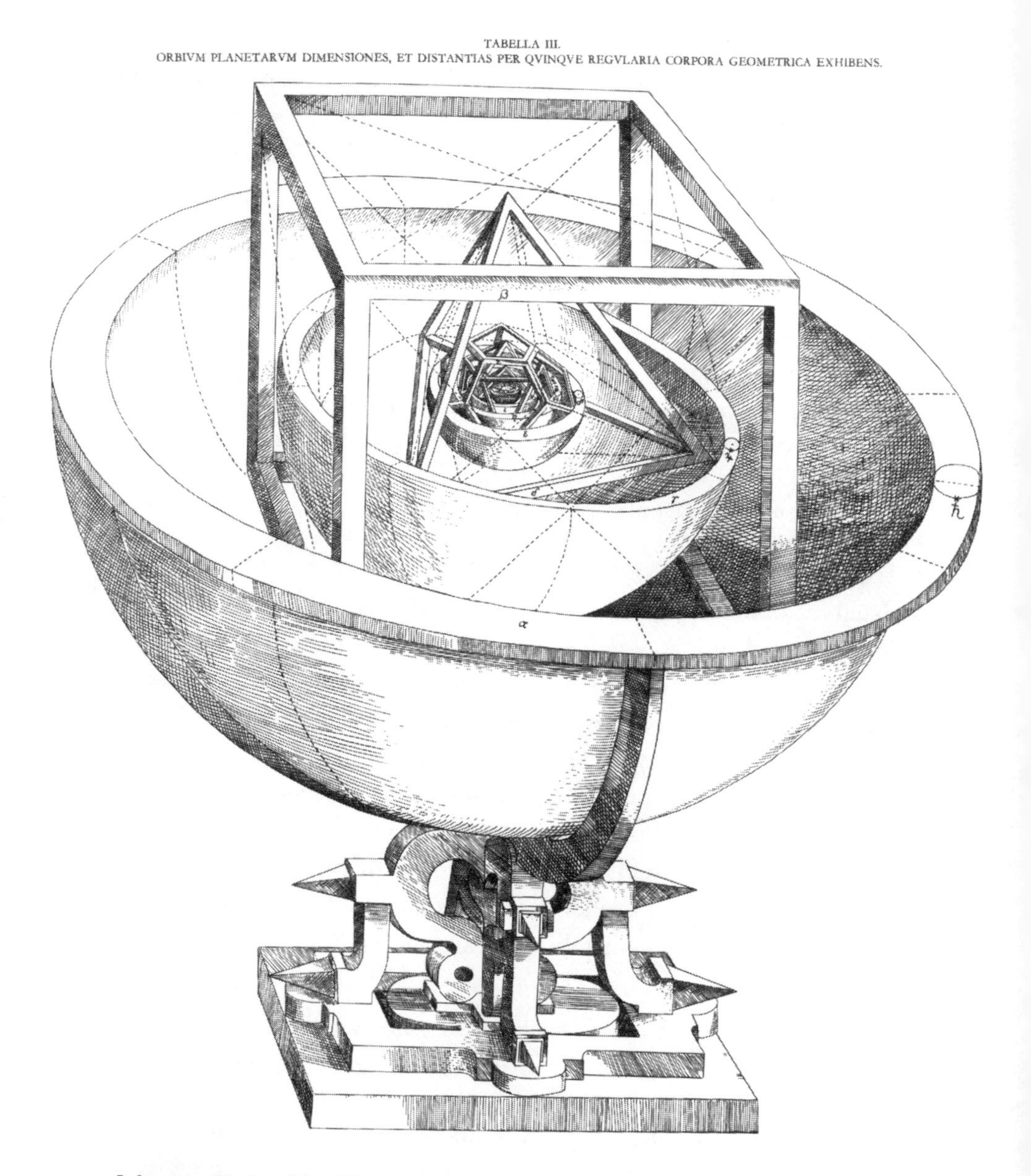

Johannes Kepler, The Planets Inscribed in the Five Regular Solids *(1596). Geometrical analysis based on Platonic and Pythagorean concepts.*

of Faraday's work, scientifically educated men and women have an altogether richer view of the world. For them, apparently empty space is full of the most intricate and beautiful patterns. So Faraday gave the world not only fresh wealth but fresh beauty."

2

NOTHING BUT FACTS

The search of Pythagorean science for elegant geometric and mathematical forms behind the appearances of the visible world stands in opposition to what can be called the materialistic tradition. Materialism assumes that the world, itself, is real. It is not the objectification of ideal forms but the sum of all the physical objects that go to make it up. The ultimate objects are atoms. The key to understanding nature is disciplined observation of objects, and the product of disciplined observation is fact.

If this is true, the development of science depends on a proper scientific method, and although Sir Francis Bacon was by no means the only Renaissance philosopher to discuss method in the sense of careful observation and rigorous sifting of fact from fiction, his formulation was a dominant influence on science from the time of his death in 1626 to the end of the nineteenth century.

For Bacon, that the world is a collection of physical objects is so obvious it scarcely needs to be argued. The central problem of science is not the physical world being observed but the fallibility of the observer. The mind is always playing tricks, sometimes by chance but often because of the very way it works. Information about the world enters the mind through the senses, but the senses are notoriously unreliable. They color and distort information even as they transmit it. Then there is imagination, which continuously changes what the senses report to fit its own program. Man needs facts, but facts are hard to come by. He is locked in a struggle not only with traditional error but with his own nature. As Bacon remarks in *The Advancement of Learning* (1606), "The mind of man is far from the nature of a clear and equal glass, wherein the beams of things should reflect according to their true incidence; nay, it is rather like an enchanted glass, full of superstition and imposture, if it be not delivered and reduced."

There is a strong note of asceticism in Bacon. Man is not the supreme work of Creation, the measure of all things and the image of God, as he is, for example, in Michelangelo's heroic painting of Adam touched by the finger of God in the Sistine Chapel. Instead he is an upstart crow in a natural world utterly indifferent to his existence. Being an enchanted glass, that is, a distorting mirror, his mind must be "delivered and reduced," that is, mortified, if it is to find truth.

Since this is a hostile critique of the human mind, it is in some sense an attack on that which makes people human. Poetry is an especially flagrant surrender of the mind to the vanities of imagination. It is a survival in the modern world of primitive habits of mind—of "rude times and barbarous Regions, where other learning stood excluded."

The nature of poetry is to distort the real. A metaphor joins things not joined in nature. The more unlike the things, the more striking the effect. Bacon considered the attachment of poetry to metaphors pernicious. It "doth truly refer to the Imagination, which, being not tied to the Laws of Matter, may at pleasure join that which Nature hath severed, and sever that which Nature hath joined, and so make unlawful Matches and divorces of things."

In spite of his suspicion of metaphors, Bacon cannot avoid them when explaining the tasks assigned to Reason. The metaphors are of warfare and enslavement. Reason must "deliver and reduce" the mind. Once the mind is conquered, reason eliminates its innate poetic inclinations and "buckles and bows" it to the nature of things: "Poetry . . . doth raise and erect the Mind, by submitting the shows of things to the desires of the Mind, whereas reason doth buckle and bow the Mind unto the Nature of things."

Since Bacon had no experience of sophisticated scientific instruments, he understood reality as things seen in the middle distance by the naked eye. Things are as far from poetry as you can get. The ideal Baconian dictionary of reality is a collection of pictures, one for each sort of thing that exists in the world. It follows that the ideal language is a language of nouns with one noun for each thing, so that men will speak "so many *things* almost in an equal number of *words*." There will not be one metaphor in that dictionary.

A fact is a verifiable assertion about a thing. It is the opposite of a poem. The purpose of Bacon's experimental method is to separate facts from poetry—from superstitions, prejudices, fancies, and all of the other errors that mind introduces into its image of the world.

This faith in *things* is of a piece with the hostility of seventeenth-century materialists toward geometric explanations of nature. Geometry is a product of mind. To assert that cubes and tetrahedra are real—more real, even, than things—is to succumb to the most pernicious kind of human vanity, the kind that presumes the mind can impose conditions on the world.

Reality is things. What kind of things? The answer is very small things, invisible things—atoms, in fact.

The Baconian faith in the knowability of the world blended easily in the seventeenth century with a revival of the ancient theory, derived ultimately from Leucippus and Democritus, that matter is composed of atoms. Atoms may not be observable, but they are definitely things. Force is communicated by their collisions. After colliding, they follow the neat trajectories of billiard balls.

Since invisible things cannot be observed, how do scientists know atoms are not just as much products of imagination as mystical geometry? The answer is that you can reach out and touch things. You can't reach out and touch ideas or qualities.

Vision and hearing are notoriously subject to the distortions Bacon so thoroughly documented. However, you can be deaf and blind, and if you touch something and it is hard, you know it exists. Because touch is so basic, because it seems to measure only a single aspect of the real, it seems more reliable than the other senses. To say something is "palpable" is synonymous, after all, with saying it is there. But touch, when analyzed, is based on the hardness—or "impenetrability"—of the thing touched. If you can punch it and it does not disappear it is a fact.

In a "query" at the end of his *Optics* (1717), Newton carried the idea of hardness to its logical limit, equating it with divine power: "God in the Beginning formed Matter in solid, massy, hard, impenetrable, movable Particles. . . . these primitive Particles being Solids, are incomparably harder than any porous Bodies compounded of them; even so very hard as never to wear or break in pieces; no ordinary power being able to divide what God made one in the first Creation." (Query 31)

"*Solid, massy, hard, impenetrable* . . ." There is an appeal here to feelings so primitive they may go back to the first encounters of infants with a world that makes itself evident by refusing to budge when they push against it. A few years after Newton's *Optics*, Bishop George Berkeley would argue that reality depends for its existence on being perceived. Much incensed by this attack on materialism, Dr. Samuel Johnson offered a famous rebuttal that depends for its effect entirely on the association of reality with impenetrability. He kicked a stone and remarked, "Thus I refute Berkeley."

In spite of the hardness of things, when pictures of them get inside the mind, they tend to be made into what they are not. There's the rub. People need facts, not fancies. It is not only ideas that are suspect, but qualities like color and warmth and responses like the sense of beauty. Red is a mental image; what exists in the world is light of a certain frequency. Beauty is a human emotion. A sunset is not beautiful. It is not even red. It is just there. The mind is always throwing seductive and misleading disguises over things. In Charles Dickens's *Hard Times*, Mr. Gradgrind banishes flowered carpets from his school because they accustom the pliable minds of children to false images and thus to the corrupting delights of fancy. The first paragraph of the novel is a tribute to the robust survival of Baconian asceticism two centuries after Bacon's death: " 'Now, what I want is, Facts. Teach these boys and girls nothing

```
--- YOU ARE DONE. TYPE NEXT TO GO ON AFTER ---
--- TESTING THE FUNCTIONS YOU HAVE DEFINED ---

(defun fact (n)
    (cond ((zerop n) 1)
          (t (times n (fact (sub1 n))))))

                THE LISP WINDOW

= > (trace fact)
(fact)

= > (fact 3)
1 <Enter> fact (3)
|2 <Enter> fact (2)
| 3 <Enter> fact (1)
| |4 <Enter> fact (0)
| |4 <EXIT> fact 1
| 3 <EXIT> fact 1
|2 <EXIT> fact 2
1 <EXIT> fact 6
6
```

"Call for Facts" (computer screen from Byte *magazine, April 1985).*

but Facts. Facts alone are wanted in life. Plant nothing else, and root out everything else. You can only form the minds of reasoning animals upon Facts: nothing else will ever be of any service to them.' "

This is not the whole story, and there is a touch of Pythagorean wonder in Bacon in spite of himself. But the distrust of mind, not the wonder, is what made Bacon a hero of science in the eighteenth and nineteenth centuries, particularly of sciences that depended primarily on observation and classification: geography, geology, biology, anatomy.

3

CHARLES DARWIN'S TREE OF LIFE

The culmination and—for many Victorians—the vindication of the Baconian tradition in science was Charles Darwin's *Origin of Species* (1859). Darwin acknowledges his debt to Bacon in his *Autobiography* (1876): "I worked on the true Baconian principles, and without any theory collected facts on a wholesale scale."

Wholesale is right. The book brings together twenty years of painstaking, minutely detailed observation ranging over the whole spectrum of organic life. Like Bacon, Darwin made little use of mathematics, although he had attempted (unsuccessfully) to deepen his mathematical knowledge while at Cambridge. Nor was Darwin the sort of scientist whose observations depend on instruments. His four-volume study of barnacles—*Cirripedia* (1851–54)—uses microscopy frequently, but much of

"A Thing in Middle Distance." Centerpiece designed by Prince Albert (1849).

his best work could have been written entirely on the basis of direct observation.

As soon as it was published, *The Origin of Species* was recognized as one of those books that change history. Its reception was partly a tribute to the overwhelming wealth of detail it offers in support of the theory Darwin finally worked out to hold his enormous bundle of facts together and partly a case of powder waiting for a spark. Darwin was initially criticized for giving insufficient credit to his predecessors, and the third edition of *The Origin of Species* includes a list of important moments in the earlier history of the theory of evolution. It begins with Jean-Baptiste

Lamarck, who proposed a generally evolutionary theory of biology in the *Histoire naturelle des animaux* (1815). Charles Lyell's *Principles of Geology* (1832) is not included in the list because it is not specifically evolutionary, but its analysis of the evidence of geological change over time was indispensable to Darwin. Using Lyell, he could be certain that the variations he observed among animals of the same species in the Galápagos Islands had occurred within a relatively short span of geologic time.

Another source mentioned in the list and the immediate stimulus to the publication of *The Origin of Species* was an essay by Alfred Russell Wallace entitled "On the Tendency of Varieties to Depart Indefinitely from the Original Type." Wallace sent this essay to Darwin in 1858, and it convinced Darwin that if he did not publish his own work he risked being anticipated. He acknowledges Wallace's paper in his introduction and admits in the *Autobiography* that it "contained exactly the same theory as mine." Again according to the *Autobiography*, it was Darwin's reading of Malthus that suggested, around 1838, that all species are locked in a remorseless struggle for survival.

In spite of these and other anticipations, *The Origin of Species* was an

"Galápagos Lizard." From Charles Darwin, The Voyage of the Beagle *(Collier ed., 1909). The image nicely defines the middle perspective from which Darwin preferred to view nature.*

immediate sensation. By ignoring religious dogma and wishful thinking, Darwin was able to buckle and bow his mind to the nature of things and to produce the sort of powerful, overarching concept that reveals coherence in a vast area of experience that had previously seemed chaotic.

A modern reader can see a kinship between Darwin's passionate interest in all things living, beginning with his undergraduate hobby of collecting beetles, and the outburst of nature poetry that occurred in the Romantic period.

Darwin was unaware of this kinship. In the *Autobiography* he says that "up to the age of thirty, or beyond it, poetry of many kinds, such as the works of Milton, Gray, Byron, Wordsworth, Coleridge, and Shelley . . . gave me the greatest pleasure. . . . But now for many years I cannot endure to read a line of poetry." His *Journal of the Voyage of the Beagle* is filled with appreciative comments about tropical landscape and its animals and plants, but he remarks that natural scenery "does not cause me the exquisite delight which it formerly did." He is probably contrasting his own methodical descriptions of landscape with the romanticized landscapes of writers like Byron and painters like Turner. He plays the role of Baconian ascetic collecting "without any theory . . . facts on a wholesale scale." His mind, he says (again in the *Autobiography*), has become "a kind of machine for grinding laws out of large collections of facts."

The idea that the mind is a machine that grinds laws out of facts echoes Bacon's injunction to use reason to "deliver and reduce" the imagination. The same asceticism is evident in Darwin's disparaging comments about his literary style. He believed he was writing dry scientific prose for other scientists, and John Ruskin, among others, agreed. Darwin was astounded, gratified, and a little frightened by his popular success.

No one can read Darwin today without recognizing that he was wrong about his style. As Stanley Edgar Hyman observes in *The Tangled Bank* (1962), both *The Voyage of the Beagle* and *The Origin of Species* are filled with passages that are beautiful and sensitive, whatever Darwin may have thought of them. The writing is effective precisely because it does not strain for the gingerbread opulence fashionable in mid-Victorian English prose. It has a freedom from pretense, a quality of authority, as moving as the natural descriptions in Wordsworth's *Prelude*. It is effective precisely because it stems from direct observation of the things and relationships that nature comprises. In addition to revealing a mind "buckled and bowed" to nature, it reveals a mind that has surrendered to the kaleidoscope of life around it. Consider the following comment on the life-styles of woodpeckers:

> Can a more striking instance of adaptation be given than that of a woodpecker for climbing trees and seizing insects in chinks in the bark? Yet in North America there are woodpeckers which feed largely on fruit, and others with elongated wings which chase insects on the wing. On the plains of La Plata, where hardly a tree grows, there is a woodpecker . . . which has two toes before and two behind, a long pointed tongue, pointed tail-feathers, sufficiently stiff to support the bird on a post, but not so stiff as in the typical woodpeckers, and a straight strong beak. . . . Hence this [bird] in all essential parts of its structure is a woodpecker. Even in such trifling characters as the colouring, the harsh tone of the voice, and undulatory flight, its close blood-relationship to our common woodpecker is plainly declared; yet . . . in certain large districts it does not climb trees, and it makes its nest in holes in banks! In certain other districts, however . . . this same woodpecker . . . frequents trees, and bores holes in the trunk for its nest.

Darwin was familiar with Audubon's *Birds of America*, and remarks that Audubon "is the only observer to witness the frigate-bird, which has all its four toes webbed, alight on the surface of the ocean." In spite of a possible touch of irony in this remark, the affinity between the two naturalists is striking. Darwin fixes things in the middle distance by means of words. The central device in his description of the La Plata woodpecker is detail: elongated wings, insects caught on the wing, two toes before and two behind, stiff tail, elongated beak, harsh voice, a nest in a hole in a bank. The accumulating details express close observation which is also loving observation. They create a thingly poetry, a poetry of the actual (see illustration 1 in color section).

In a similar way, Audubon fixes in images a nature that flaunts itself palpably and colorfully in the middle distance. In the process both Darwin and Audubon create an art of the actual.

A year before the *Origin of Species*, Oliver Wendell Holmes published "The Chambered Nautilus." It is a poem that attempts to fix a thing that is out there in the middle distance in verse:

Year after year behold the silent toil
That spread his lustrous coil;
Still as the spiral grew,
He left the past year's dwelling for the new,
Stole with soft step its shining archway through, built up its idle door,
Stretched in his last-found home, and knew the old no more.

Here, instead of the scientist becoming poet, the poet becomes a scientist. The problem is that the poem cannot forget it is art. It is more clumsy, finally, than Darwin's description of the La Plata woodpecker. Closer to Darwin are the photographs of Mathew Brady, the histories of Ranke and Burckhardt, and the novels of Balzac, George Eliot, and Turgenev.

Feeling is usually implicit in Darwin's prose but repressed. Facts are facts and poetry is poetry. Occasionally, however, Darwin allowed his

"Finches from Galápagos Archipelago." Variations on the Geospiza, from Charles Darwin, The Voyage of the Beagle *(Collier ed., 1909).*

feelings to bubble to the surface. The closing paragraph of *The Origin of Species* is a case in point. It describes a scene,

> . . . clothed with many plants of many kinds, with birds singing on the bushes, with various insects flitting about, and with worms crawling through the damp earth, and . . . these elaborately constructed forms, so different from each other, and dependent upon each other in so complex a manner, have all been produced by laws acting around us. . . . Thus, from the war of nature, from famine and death, the most exalted object which we are capable of conceiving, namely, the production of the higher animals, directly follows.

No passage is more obviously dominated by aesthetic feeling than Darwin's description of the variety of species created by the struggle for existence. The idea of the struggle is central to *The Origin of Species*. It involves a paradox that fascinated Darwin. Out of a silent but deadly struggle comes the infinitely varied and exotically beautiful mosaic of life:

> How have all these exquisite adaptations of one part of the organization to another part, and to the conditions of life, and of one organic being to another being, been perfected? We see these beautiful co-adaptations most plainly in the woodpecker and the mistletoe; and only a little less plainly in the humblest parasite which clings to the hairs of a quadruped or feathers of a bird; in the structure of the beetle which dives through the water; in the plumed seed which is wafted by the gentlest breeze; in short we see beautiful adaptations everywhere and in every part of the organic world.

Exquisite, *perfected*, *beautiful*, *humblest*, *plumed*, *gentlest*. The world described by these adjectives is not cold, alien, or indifferent. It is a work of art. Nor is Darwin's prose the dispassionate, dry prose of a treatise devoted only to facts. Because it is the work of a naturalist, it pays close attention to detail. The parts are there because they are there in nature in the middle distance: the woodpecker, the mistletoe, the parasite clinging to the quadruped, the feathers of the bird, the water beetle, the plumed seed. They illustrate the harmonious relations created by the struggle for survival—"co-adaptation" is Darwin's word. The prose enacts these harmonies through elegantly controlled rhythms.

Darwin's language invites the reader to share experience as well as to understand it. *Exquisite*, *beautiful*, and *gentle* orient him emotionally at the same time that his attention is focused on the objects that give rise to the emotion—mistletoe, parasite, water beetle, plumed seed.

The passage flatly contradicts Darwin's statement in the *Autobiography* that his artistic sensitivity had atrophied by the time he was thirty. That he thought it did shows only that he believed with his contemporaries that science is science and art is art. The problem was in his psyche, not his prose. The tradition that science should be dispassionate and practical, that it is a kind of servitude to nature that demands the banishment of the humanity of the observer, prevented him from understanding that he was, in fact, responding to nature aesthetically and communicating that response in remarkably poetic prose. There is no detectable difference in this passage between a hypothetical figure labeled "scientific observer" and another hypothetical figure named "literary artist."

The most striking example of Darwin's artistry occurs in the "summary" of Chapter 3. The passage deals explicitly with the tragic implications of natural selection. It is a sustained meditation on a single image. The image—the Tree of Life—is practical because the branching limbs are a vivid representation of the branching pattern of evolution. However, the image is also mythic, an archetype familiar from Genesis and also from Egyptian, Buddhist, Greek, and other sources. In mythology, the Tree of Life connects the underworld and the heavens. It is the axis on which the spheres turn and the path along which creatures from the invisible world visit and take leave of earth. It is an ever-green symbol of fertility, bearing fruit in winter. It is the wood of the Cross on which God dies and the wood reborn that announces the return of life by sending out new branches in the spring. All of this symbolism is familiar from studies of archetypal and primitive imagery. Behind it is what Rudolf Otto calls, in *The Idea of the Holy*, the terrifying and fascinating mystery of things: *mysterium tremendum et fascinans*.

It is surprising to find a scientist, particularly a preeminent Victorian scientist and a self-avowed disciple of Bacon, using an archetypal image. Yet Darwin's elaboration is both sensitive and remarkably full. Central to it is the paradox of life in death, and throughout, one senses the hovering presence of the *mysterium tremendum et fascinans*:

The affinities of all the beings of the same class have sometimes been represented by a great tree. I believe this simile largely speaks the truth. The green and budding twigs may represent existing species; and those produced during former years may represent the long succession of extinct species. At each period of growth all the growing twigs have tried to branch out on all sides, and to overtop and kill the surrounding twigs and branches, in the same manner as species and groups of species have at all times overmastered other species in the great battle for life. . . . Of the many twigs which flourished when the tree was a mere bush, only two or three, now grown into great branches, yet survive and bear the other branches; so with the

Tree of Life. *An eighteenth-century counterpane design based on this archetypal motif.*

> species which lived during long-past geological periods, very few have left living . . . descendants.
>
> From the first growth of the tree, many a limb and branch has decayed and dropped off; and all these fallen branches of various sizes may represent those whole orders, families, and genera which have now no living representatives, and which are known to us only in a fossil state. As we here and there see a thin, straggling branch springing from a fork low down in a tree, and which by some chance has been favored and is still alive on its summit, so we occasionally see an animal like the Ornithorhynchus or Lepidosiren, which in some small degree connects by its affinities two large branches of life, and which has apparently been saved from fatal competition by having inhabited a protected station. As buds give rise by growth to fresh buds, and these, if vigorous, branch out and overtop on all sides many a feebler branch, so by generation I believe it has been with the great Tree of Life, which fills with its dead and broken branches the crust of the earth, and covers the surface with its ever-branching and beautiful ramifications.

Darwin's music here is stately and somber. The central image is established at the beginning: a great tree green at the top but filled with dead branches beneath the crown. The passage becomes an elegy for all the orders of life that have perished since the tree began. Words suggesting death crowd the sentences: *overtopped*, *kill*, *the great battle for life*, *decayed*, *dropped off*, *fallen*, *no living representative*, *straggling branch*, *fatal competition*. As the passage moves toward its conclusion, a change, a kind of reversal, can be felt. Words suggesting life become more frequent: *alive*, *life*, *saved*, *fresh buds*, *vigorous*. The final sentence restates the central paradox in a contrast between universal desolation—"dead and broken branches [filling] the crust of the earth"—with images of eternal fertility—"ever-branching and beautiful ramifications."

In spite of the poetic qualities of *The Origin of Species*, the idea of science as the dispassionate observation of things is central to the Darwinian moment. Observation reveals truth; and once revealed, truth can be generalized.

The truths discovered by Darwin were applied almost immediately to sociology and political science. Herbert Spencer had coined the phrase "survival of the fittest" in 1852 in an article on the pressures caused by population growth entitled "A Theory of Population." Buttressed by the prestige of *The Origin of Species*, the concept of the survival of the fittest was used to justify laissez-faire capitalism. Andrew Carnegie remarked in 1900, "A struggle is inevitable [in society] and it is a question of the survival of the fittest." John D. Rockefeller added, "The growth of a large business is merely the survival of the fittest." Capitalism enables the strong to survive while the weak are destroyed. Socialism, conversely, protects the weak and frustrates the strong. Marx turned over the coin: socialism is a later and therefore a higher product of evolution than bourgeois capitalism. Being superior, it will replace capitalism as surely as warm-blooded mammals replaced dinosaurs.

Darwin also influenced cultural thought. To say this is to say that he changed not only the way the real was managed but the way it was imagined. The writing of history became evolutionary—so much so that historians often assumed an evolutionary model and tailored their facts to fit. The histories of political systems, national economies, technologies, machinery, literary genres, philosophical systems, and even styles of dress were presented as examples of evolution, usually interpreted to mean examples of progress from simple to complex, with simple considered good, and complex better.

And, of course, Darwin's theories were both attacked and supported in the name of religion. Adam Sedgwick, professor of geology at Cambridge, began the long history of attacks on Darwin when he wrote in "Objections to Mr. Darwin's Theory of the Origin of Species" (1860): "I cannot conclude without expressing my detestation of the theory, because of its unflinching materialism." Among the sins for which Darwin was most bitterly attacked was his argument that species are constantly coming into existence and dying, an argument that contradicts the fundamentalist reading of Genesis. He was also attacked for suggesting that struggle, including violent struggle, is ultimately beneficial, and that, by implication, the meek will not inherit the earth. Finally, he was attacked for suggesting that man is an animal sharing a common ancestor with the apes, an idea that is implicit in *The Origin of Species* and stated unequivocally in *The Descent of Man* (1871).

Darwin's conclusion to *The Origin of Species* is a summary of his vision. It has a strong emotional coloring even in its initial form. Perhaps because of the attacks, Darwin added the phrase "by the Creator" to the first revised (1860) and later editions of the book: "There is a grandeur in this view of life, with its several powers, having been originally breathed by the Creator into a few forms or into one; and that, whilst this planet has gone cycling on according to the fixed law of gravity, from so simple a beginning endless forms most beautiful and most wonderful have been, and are being evolved."

Whether the reference to God represents Darwin's personal view of religion is outside the scope of the present discussion. Probably it did not. At any rate the notion that God is revealed in evolution remains powerfully attractive today both to biologists and, as shown by Teilhard de Chardin's *The Phenomenon of Man* (1955), to those attempting to formulate a scientific theology. More generally, in spite of his literary disclaimers, Darwin initiated a whole genre of writing, typified by the work today of Bertel Bager, Lewis Thomas, and Annie Dillard, which dwells on the intricate beauties of natural design.

Many of the applications of Darwin's ideas were, however, patently strained from the beginning. Time revealed the inadequacies of others. Social Darwinism is studied in history classes but is no longer a viable political creed. Evolutionary histories of this and that are still being written, but the approach has been shown to be seriously misleading in many applications. More fundamental, by the middle of the twentieth century Baconian empiricism was no longer adequate to the idea of nature that science had developed. Einstein and Heisenberg made it clear that mind and nature—subject and object—are involved in each other and not separate empires. An objective world that can be "observed" and "understood" if only the imagination can be held in check simply does not exist. Facts are not observations "collected . . . on a wholesale scale." They are knots in a net.

4

D'ARCY THOMPSON'S STREAMLINED FISH

D'Arcy Wentworth Thompson (1860–1948) was everything in biology that Darwin was not. *On Growth and Form* (first ed., 1917) is an attempt to create a new basis for biology by placing it on a mathematical foundation. Instead of being a collection of things in the middle distance, his nature is the materialization of mathematics. He remarks in his introduction, "My sole purpose is to correlate with mathematical statement and physical law certain of the simpler outward phenomena of organic growth and structure or form." The interest in the mathematics of biology was prophetic. Within five years after Thompson's death, Francis H. Crick and James D. Watson had shown that the double helix is the basis of the DNA molecule, which, in turn, is the basis of life.

A common criticism of Bacon's philosophy of science, as well as Dar-

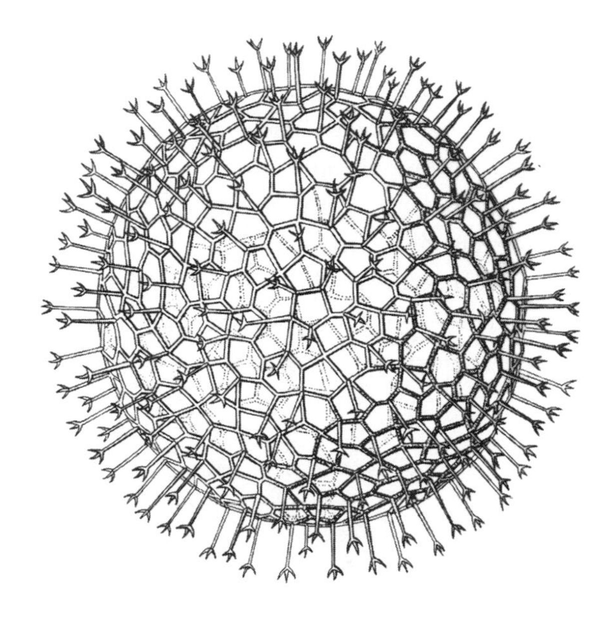

Nassellarian. From D'Arcy Thompson, Of Growth and Form *(1917). The skeleton exhibits regularities that fascinated Thompson because of their mathematical formalism.*

win's, is that it ignores mathematics. The criticism is valid. Bacon published grandiose schemes for the reform of knowledge, but Galileo established the basis for the mathematical analysis of accelerated motion and Descartes laid the foundations of analytic geometry. As we have seen, Darwin was well aware of his limited grasp of mathematics. To move from Darwin to D'Arcy Thompson is to move from the poetry of things to the poetry of the numbers that underlie things. It is to move, in other words, from things visible to things invisible.

Karl Gauss, James Clerk Maxwell, Ernest Rutherford, Josiah Gibbs, Jean-Baptiste Fourier, Hermann Helmholtz, Georg Cantor, and a host of other brilliant nineteenth-century figures shared Thompson's appreciation of the power of mathematics. All were far more accomplished mathematicians. Thompson seldom goes beyond elementary mathematical concepts. He notes, for example, that for objects of equal density, weight increases as the cube of linear dimension rather than in proportion to it. To take the simplest case, if a box is two inches on each side and is filled with a substance weighing one ounce per cubic inch, the total weight of the substance will be eight ounces, or two cubed. Double the side of the box to four inches, and the total weight contained will be sixty-four ounces, or four cubed.

The observation is commonplace, almost trivial. However, Thompson's application is not. Assuming that in animals of the same species, proportions tend to be preserved regardless of size, he concludes that the length of the limbs of land animals of the same species will, in general, increase in proportion to the cube root of their body weight. This leads to calculations of the probable lengths of the legs of birds of the same species but different weights.

Insects breathe by diffusion of oxygen through capillary tubes extending from shell to bloodstream. Their maximum size is fixed by the fact that if they become very large their breathing apparatus ceases to function.

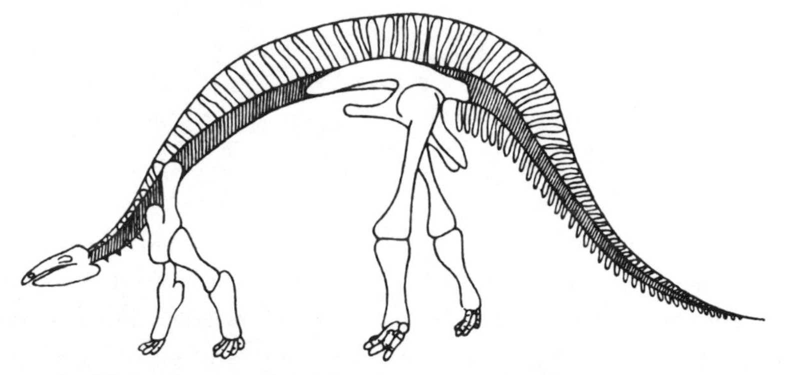

Dinosaur backbone showing arch-like adaptation to load-bearing. From D'Arcy Thompson, Of Growth and Form *(1917).*

Load-bearing horizontals must be supported, and the support structures tend to be mirror images of the stresses to which the structures are subject—hence the shapes of dinosaur skeletons and the shape of the human femur. Much speculation and not a little mumbo-jumbo has been stimulated by the architecture of honeycombs. Thompson observes that when circles are close-packed they deform, if flexible, into hexagons—hence honeycombs. The hexagons created by bees are not wondrous examples of insect intelligence or instances of the divinity infused into even the humblest creatures. Honeycombs are the way they are because any other method of making them would use more space. Square honeycombs would also use more wax, and octagonal or round ones would leave holes. Spheres tend to deform when close-packed into rhombic dodecahedra or complex fourteen-hedra—hence two common configurations of close-packed tissue cells.

The equiangular spiral (so-called by Descartes) increases in volume without changing shape—hence the chambered nautilus. If you don't shed your skin like a cicada when you grow and your skin won't grow with you like human skin or the shell of an oyster, then some sort of a spiral is the only way to go. The "stately mansions" that Oliver Wendell Holmes saw when he examined the chambered nautilus might better be described as mass-produced, low-cost portable housing. Streamlining—the term was apparently invented by Thompson—reduces turbulence to a minimum and thus maximizes swimming efficiency. It is an adaptation to the way that water flows across different surfaces. You might say that the fish have been sculpted by the water. The results, at any rate, are elegant mathematical forms that anticipate the forms of ships' hulls and airfoils.

Mathematics can help in the classification of species of animals. Bones and body shapes that appear unrelated on casual inspection can be seen to have a family resemblance when they are drawn on graphs and the graphs are systematically distorted. The illustrations of this principle in Thompson's chapter on comparative morphology show a surrealistic assortment of enlargements, reductions, local bulgings and crimpings, and upward and downward funnelings. They anticipate by half a century several of the strategies of computer-assisted design.

Fish with regular deformations to show relations among apparently different species. From D'Arcy Thompson, Of Growth and Form *(1917).*

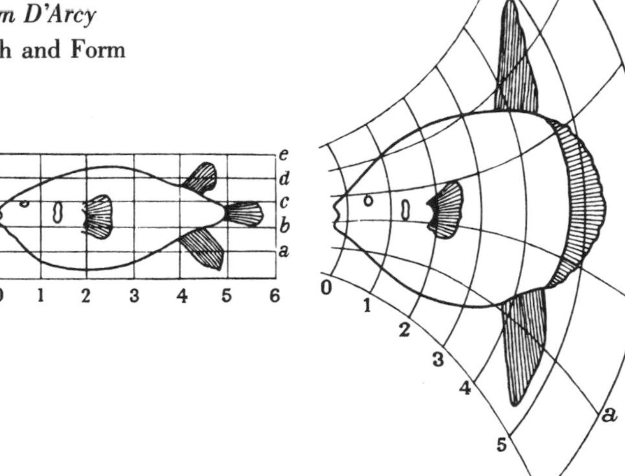

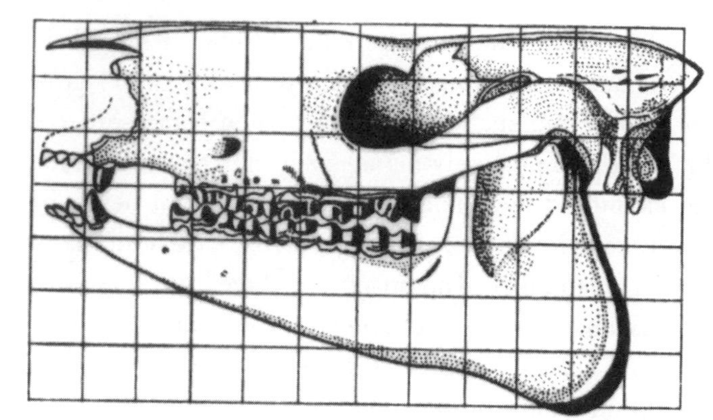

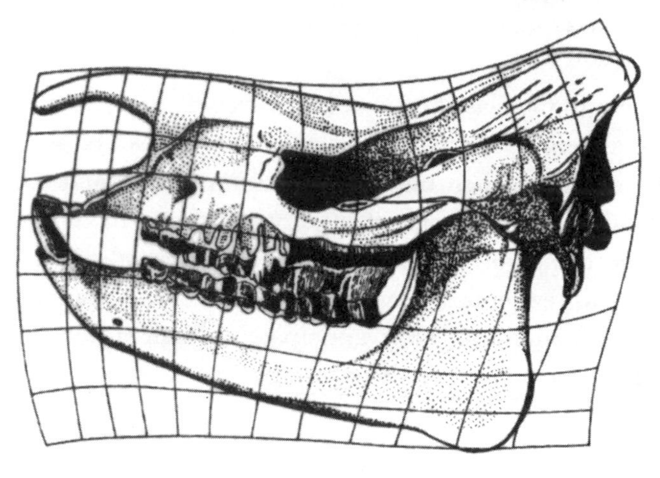

Regular deformation of animal skulls. From D'Arcy Thompson, Of Growth and Form *(1917).*

The application of mathematics to life may not seem innovative today, but Thompson clearly felt in 1917 that it was a bold, even a risky, venture. It seems to violate an unspoken feeling that life is beyond quantification. He wrote, "To treat the living body as a mechanism was repugnant, and seemed even ludicrous, to Pascal; and Goethe, lover of nature as he was, ruled mathematics out of place in natural history." Thompson added that when the zoologist meets a simple geometrical shape in a living organism, "He is prone of old habit to believe that after all, it is something more than a spiral or a sphere, and that in this 'something more' there lies what neither mathematics nor physics can explain. . . . In short, he is deeply reluctant to compare the living with the dead."

The idea that life is "something more" is a form of vitalism. It has been in constant retreat since the Middle Ages as advances in chemistry have shown the degree to which life processes are continuous with the nonliving reactions of organic chemistry. Yet vitalism is a hardy plant. Few scientists will argue in the late twentieth century that life itself is an inexplicable mystery, but arguments persist that human mental processes are irreducibly mysterious and can never be duplicated by machines. We will return to this issue when we consider machine intelligence.

Thompson obviously hoped to shelter *On Growth and Form* from the violent attacks that were made on Darwin. His analysis of the mathematics of life looks timid by comparison to *The Origin of Species*, but that is only on the surface and Thompson knew it.

Darwin's use of the Tree of Life metaphor suggests a powerful lingering sense of nature as sacramental mystery. Thompson's religion is far more rarefied. He has little sense of reverence for nature as direct experience—as a thing in the middle distance.

Thompson's surrealistic and distorting graphs of fish are applied mathematics on the surface. At a deeper level, they suggest that beyond his scientific interest in revealing hidden relations among species, he enjoys demonstrating the superiority of mathematics to the supposedly sacrosanct shapes of organic nature. The implication is that nature is clay and mathematics the sculptor. This does not mean that Thompson had no religious feeling for what he examined. It only means that like earlier Pythagoreans, he felt the divine to be manifested in geometry and number, not in matter. Matter is appearance, number is reality. The hidden agenda of the distorting graphs is to humble matter, to teach it a proper subordination to geometry and number.

On Growth and Form might have been an unreadable collection of formulas, graphs, temperature and pressure tables, stress measurements, and the like. It is not. The mathematical regularities that Thompson discovered throughout the living world seemed astonishing and delightful to him.

The Epilogue of *On Growth and Form* is a majestic expression of the mixture of analysis and Pythagorean wonder that pervades the book:

> That I am no skilled mathematician I have little need to confess. I am "advanced in these enquiries no farther than the threshold"; but something of the beauty and use of mathematics I think I am able to understand. I know that in the study of material things, number, order, and position are the threefold clue to exact knowledge; that

> these three, in the mathematician's hands, furnish the "first outlines for a sketch of the Universe"; that by square and circle we are helped, like Emile Verhaeren's carpenter, to conceive *"Les lois indubitables et fécondes qui sont la règle et la clarté du monde"* [". . . the indubitable and fertile laws that give order and clarity to the world"].
>
> For the harmony of the world is made manifest in Form and Number, and the heart and soul and all the poetry of Natural Philosophy are embodied in the concept of mathematical beauty. A greater than Verhaeren had this in mind when he told of "the golden compasses prepared In God's eternal store." A greater than Milton had magnified the theme and glorified Him "that sitteth upon the circle of the earth, saying: 'He had measured the waters in the hollow of his hand, and meted out heaven with the span, and comprehended the dust of the earth in a measure. . . .' "
>
> Not only the movements of the heavenly host must be determined by observation and elucidated by mathematics, but whatsoever else can be expressed by number and defined by natural law. This is the teaching of Plato and Pythagoras, and the message of Greek wisdom to mankind. So the living and the dead, things animate and inanimate, we dwellers in the world and the world wherein we dwell . . . are bound alike by physical and mathematical law. "Coterminous with space and coeval with time is the kingdom of Mathematics; within this range her dominion is supreme."

This is a summary of the main thesis of *On Growth and Form* and a hymn to the comely beauty of the world. Thompson quotes poetry—Verhaeren, Milton, and Scripture—to find words adequate to his vision, but his own language is equally poetic. His reality is an eternal sequence of perfect forms in which change is an illusion and the many are simply disguises assumed by the One. The bewilderingly varied phenomena of nature that so delighted Darwin—the skeletons of diatoms, the cells of honeycombs, the shells of the Nautilus, the skulls of horses—are reduced by proper analysis to the dodecahedron and hexagon and logarithmic spiral and truss. He quotes Galileo to point the contrast between true analysis and "mere description"—that is, between numbers on the one hand and things and facts on the other: "The mathematical definition of 'form' has a quality of precision which was quite lacking in our earlier stage of mere description. . . . We are brought by means of it in touch with Galileo's aphorism (as old as Plato, as old as Pythagoras, as old perhaps as the wisdom of the Egyptians) that 'the Book of Nature is written in mathematical terms.' "

Thompson believed in 1917 that he had embarked on a journey even more radical than that of physics. Mathematics had been a part of physics since Pythagoras, but the application of mathematics to life was, Thompson felt, something new. By extending its empire to living organisms—and we have already seen that he was uneasy about the step—Thompson made it a universal category. There is nothing left to hold out against it. A revolution has been won.

Although Darwin considered his nature material and objective, it is saturated with human motives. Life is made possible only by death. The cost of the beautiful and ever-ramifying branches on the Tree of Life is the graveyard of species that litter the earth around its base. To regard the struggle for existence as tragic, however, is logical nonsense. It is a prime example of the pathetic fallacy—hence, in spite of Darwin's allegiance to Bacon, an instance of what Bacon considered the besetting human error: "submitting the shows of things to the desires of the Mind." The million plumed seeds that die in order that one may live have no sense of their own tragedy. It is the human observer who imposes this sense on the order of things.

To call death—even the death of seeds—tragic is to call life a blessing. But to say that the death of seeds is, purely and simply, the death of seeds is to suggest that life and death have equal meaning or no meaning at all. They are phases of matter, sides of a coin, a yin and a yang. The emotional intensity of Darwin's description of the struggle for survival suggests strongly that he was psychologically unable to admit this possibility.

Thompson was. Mathematics has no place for the pathetic fallacy. When viewed in terms of the category of mathematics the world retains its dazzling beauty, but it is not tragic. Instead of clothing the nonhuman world with human sentiments, mathematics seems to do the reverse. It seems to strip away the claim of life to being somehow unique by treating it as continuous with the inorganic world. A sphere is a sphere whether it is materialized in a raindrop or a human tear. A truss structure is a response to loads placed on it whether it is an iron bridge or a human femur. A honeycomb is the cheapest and most labor-efficient way that geometry offers to make cells for bee larvae. The inwardness of the living organism is continuous with the outwardness of inert matter: "We dwellers in the world and the world wherein we dwell," Thompson wrote, "are bound alike by physical and mathematical law."

The word "seems" in the preceding paragraph is used advisedly because the moment in the development of scientific culture represented by D'Arcy Thompson confronts us with a paradox. Mathematics is not "in the world" in the sense that Bacon's things and Newton's atoms and Darwin's woodpeckers and water beetles are in the world. It may be in the mind of God. If so, mathematics is the true ground of reality. On the other hand, if God does not exist—or if God exists but is not interested in mathematics—then mathematics is a product of the human mind. It is not the ground of the real but a mask imposed on the real by man. It is a fiction so skillfully devised that thinkers as remarkable as Galileo and D'Arcy Thompson become convinced that it has a basis in the world or—more precisely—*is* the basis of the world.

The same paradox is evident in another aspect of modern science. It has moved away from observations at the middle distance where things look like things to observations of the very small and the very distant, and the instruments it uses to observe often yield images quite different from those created by the five unaided senses.

The effect of seeing nature from these perspectives is apparently to dehumanize it. Things no longer look like things. The images that the instruments produce have more kinship with each other than with what they "are." An enlarged microphotograph of a cancer cell looks like an abstract painting, or, to the untrained and unwary eye, much like a microphotograph of crystals embedded in an alloy or like geological formations in radar images made from a satellite. An infrared view of a distant landscape may glow like an Impressionist canvas; an ultraviolet view will be different but equally lovely (see photo 2 in color section).

All of these versions of nature are "real." In fact, from Thompson's point of view they are more real than nature observed from a middle distance because they show structure. Seeing structure is a first step to seeing the mathematical basis of structure.

Considered thoughtfully, the images produced by the instruments are not "dehumanized" versions of nature. The reverse is true. They are radically human. The instruments were made by men, not by God or nature, and it is only through their mediation that the images can appear.

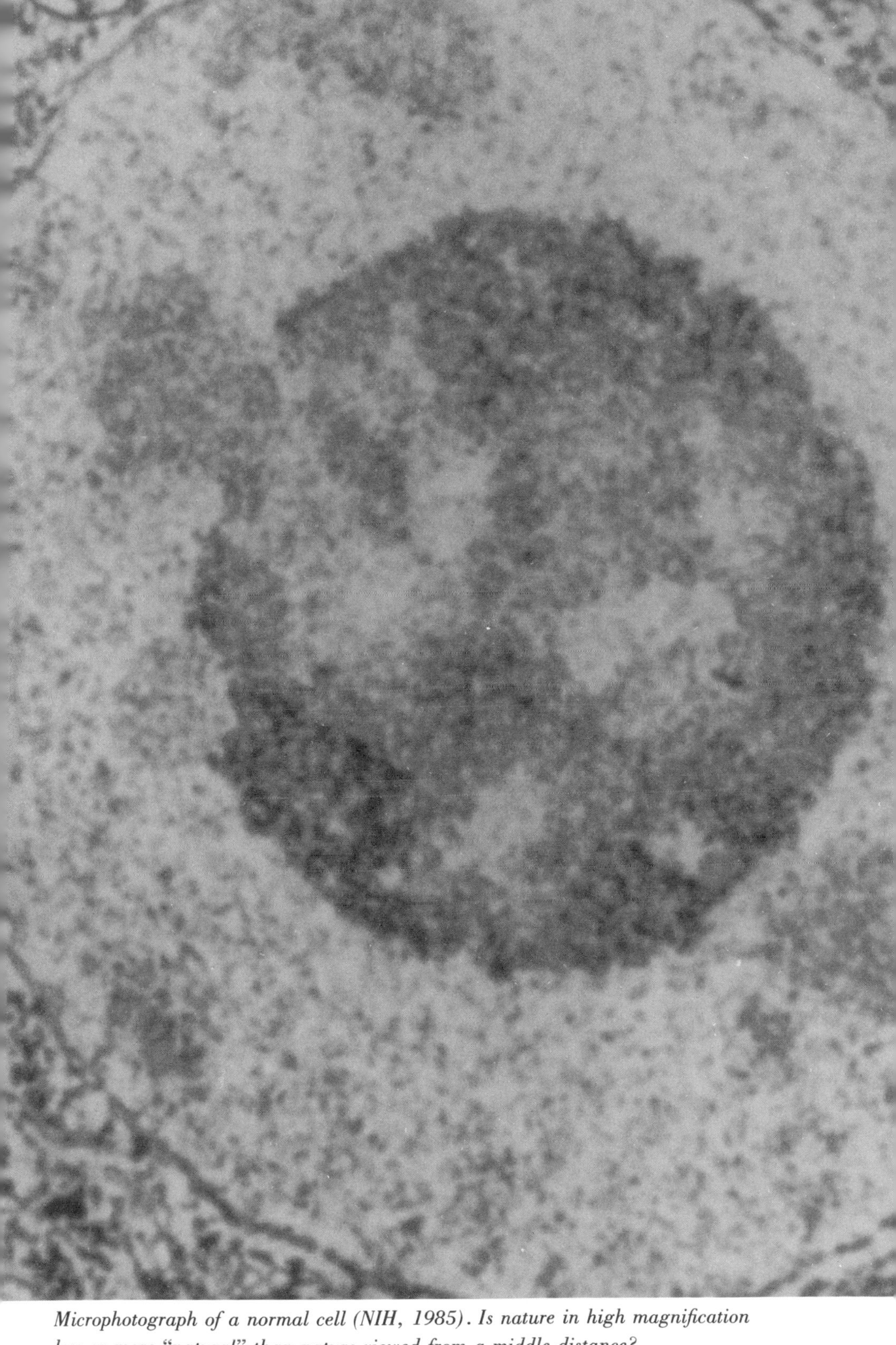

Microphotograph of a normal cell (NIH, 1985). Is nature in high magnification less or more "natural" than nature viewed from a middle distance?

The nature they reveal is not an alien "other," whether beautiful or terrible or both. It is, rather, a wedding of the human spirit with things.

At the same time, the image of nature in the sense of things seen from the middle distance has disappeared. The contrast between Darwin and Thompson shows that as the image dissolves, the sense of tragedy—of human drama with human consequences—is replaced by generalized appreciation of the beauty of form. Thompson's vision has no place for the sense of tragedy. The sense of detail, the concern for the particularity of each object, that is so powerful in *The Origin of Species* is transformed into delight in the discovery of mathematical regularities replicated across the whole range of life.

Inevitably Thompson is forced by his assumptions into direct confrontation with Darwin. Darwin was nothing if not historical. Evolution takes place over an immense span of time that dwarfs human history. It also has a direction which tends to be up, an ascent from simple to complex, which Darwin unhesitatingly invested with ethical significance. To move up on the scale of complexity is to move from good to better. Millions die but the fittest survive. Biological species gradually improve, and the triumphant climax of the process is man. This is the consolation that finally allows Darwin to come to terms with the struggle for survival. Whether or not he believed his own statement, Darwin announced that the wondrous benefits of natural selection testify to the presence of the Creator at the beginning of the process.

For Thompson, biological time is a meaningless concept:

> In the order of physical and mathematical complexity there is no question of the sequence of historic time. The forces that bring about the sphere, the cylinder or the ellipsoid are the same yesterday and to-morrow. A snow-crystal is the same today as when the first snows fell. The physical forces which mould the forms of *Orbulina*, of *Astrorhiza*, of *Lagena* or of *Nodosaria* to-day were still the same . . . in that yesterday which we call the Cretaceous epoch; or, for aught we know, throughout all that duration of time which is marked, but not measured, by the geological record.

This is much more than a quibble. It is a challenge to the conceptual basis of the theory of evolution.

Thompson's biology is concerned with functional relationships rather than ancestries. The search of evolutionary biologists for "the blood relationships of things living and the pedigrees of things dead and gone" is harmless but irrelevant. It must give way in a proper system of biology to the analysis of "fundamental properties" and "unchanging laws, of matter and energy."

Having rejected evolution Thompson also rejects Darwinian progress:

> In the end and upshot, it seems to me by no means certain . . . that the concept of continuous historical evolution must necessarily, or may safely and legitimately, be employed. That things not only alter but improve is an article of faith, and the boldest of evolutionary conceptions. How far it were true were very hard to say; but I for one imagine that a pterodactyl flew no less well than does an albatross, and that the Old Red Sandstone fishes swam as well and as easily as the fishes of our own seas.

If this position is accepted, the Creator poised so imposingly on the platform of Darwin's concluding paragraph looks a little silly, like a magician with a hat but no rabbit.

The invisible web of forces that shape natural organisms in *On Growth and Form* has no history because it is timeless and no home because it is universal. The living and the dead, the dwellers in the world and the world in which they dwell, are to the mathematician indifferent. Those who find evidence of a divine plan in the numbers of living forms may do so. This is the path of Pythagorean mystics and religiously inclined modern scientists. Thompson's book shows he was a little of both.

5

ALICE'S ANOMALIES

The problem of the world seen from different perspectives is both familiar and profound. To someone wearing dark glasses the world looks dark, but it is not necessarily dark. Take off the glasses and the world is flooded with intolerable light. One kind of mathematics reveals one kind of pattern; another reveals another. The brilliance of the world becomes yellow with yellow lenses, blue with blue lenses.

On a Mercator projection of the world, America looks different from America on a Dymaxion projection, even though it is the same America. The two mapping systems are masks imposed on the world by men. Which map is right? The first answer is that both are wrong in the sense that they misrepresent the truth about the world. The world is round, not flat.

Their wrongness is precisely what makes them useful. A flat map is much easier to use than a globe and also cheap enough so that you can mark it up, stick pins in it, and throw it away when it gets dog-eared. Since both maps are useful, the second answer is that the right map is the one that tells you the lies you need to know.

To say that human knowledge is always tangled with fiction does not imply an end of human response to nature. It places man in a country he creates partly with his own mind. In this country he is surrounded by brilliant, fantastic, wildly distorted images of himself. Is there a god behind the mask? There is no way of knowing. Hence the third moment of modern science, the authentically modern moment, the moment of reality as game.

Wallace Stevens calls the games the mind plays with the world "necessary fictions." The mind cannot get along with them, but it cannot get along without them either. They organize experience just as religion, mythology, and tradition organize it. They are the preconditions of knowledge. This is frustrating and man forever struggles against it by seeking a supreme truth or, as Stevens has it, a supreme fiction. His poet mutters:

> To discover winter and know it well, to find,
> Not to impose, not to have reasoned at all,
> Out of nothing to have come on major weather,
>
> It is possible, possible, possible. It must
> Be possible. It must be that in time
> The real will from its crude beginnings come.

In fact, it is impossible. We know nothing that has not been invested with ourselves. Not only have we learned this from Kant; we learn it from modern physics, too. Stevens knows it from his own experience as a poet trying to see the world and realizing again and again that whatever he sees is changed by the act of being seen. It is the same problem, really, that troubled Bacon when he contemplated the deceptions of imagination. Unlike Bacon, Stevens knows there is no way out. Toward the end of the poem he tries to come to terms with this aspect of the human condition. The world is a fiction, but perhaps we will eventually discover that the fiction makes sense:

. . . Yes, that.

They will get it straight one day at the Sorbonne.
We shall return at twilight from the lecture
Pleased that the irrational is rational.

Between the publication of Newton's *Principia Mathematica* in 1687 and Darwin's *Origin of Species* in 1859, science believed it could present man with truth. This was a fantasy. Science still challenges religion and mythology and tradition—and there are still people who resent the challenge—but it no longer promises to replace them with truth, only with necessary fictions. In *Masks of the Universe* (1984), Edward Harrison suggests that all of the cosmologies that men have produced since the earliest myths are so many masks covering a face that will never be seen. The world is always somewhere behind its masks.

Around 1550, space stopped being spherical and became continuous and infinite. Around 1920, it stopped being infinite and continuous and became, in one formulation, finite but unbounded—not spherical, really, but emphatically not continuous. Today astronomers suggest that it may be as littered with local funnel-shaped gravitational irregularities as a street full of potholes, and it is probably pocked with open drains called black holes. Squirming spaghettilike strings may arch across it and warp its shape like the laces of an enormous corset. This morning the world may be a rhombic dodecahedron; by noon it may well be a Möbius strip. If it is always hiding behind masks, one thing is certain: the masks are getting stranger and stranger.

Friedrich Schiller, poet and longtime friend of Goethe, argued that the human urge to create is produced by the play impulse, the *Spieltrieb*. Play and work both require effort—anybody who has stayed up all night playing poker knows how much energy is needed even for a sedentary game—and games have rules. Rules are essential because without them there would be nothing to play. But they are only rules. They are retained as long as they are useful or entertaining. They can be changed to make things more exciting. Or the game can be changed. When things get boring, you put the dice back in the drawer and set the table for chess.

To move from D'Arcy Thompson's majestic and classical meditation on the comely order of things to the world of modern physics is to move from a science that assumes the existence of absolutes to a science that is provisional, relative, paradoxical, playful, ebullient. Whatever else he did or did not do, Thompson believed he was presenting Truth. To the degree that modern science has accepted the idea of necessary fictions, it is a science liberated from the need to present Truth, hence a science that is radically playful.

"To the degree" is an important qualification. Not all scientists are ready to buy the idea of necessary fictions. Confronted with the eccentricities of quantum mechanics, Einstein was distressed. At heart he was a materialist. There must be something out there, he felt, to account for the fact that we observe it. Einstein was more like a chess player than a crap shooter. Werner Heisenberg, himself one of the pioneers in quantum theory, recalls in *Physics and Beyond*:

> Einstein had devoted his life to probing into that objective world of physical processes which runs its course in time and space, independent of us, according to firm laws. . . . And now it was being asserted that, on the atomic scale, this objective world of time and space did not exist. . . . Einstein was not prepared to let us do what, to him, amounted to pulling the ground from under his feet. . . . "God does not throw dice" was his unshakable principle, one that he would not allow anybody to challenge. To which [Niels] Bohr could only counter: "Nor is it our business to prescribe to God how He should run the world."

In spite of Einstein, it is fair to say that the dominant movement of twentieth-century theoretical science, especially mathematics, physics, and cosmology, has been away from certainties and toward masks and games. This is undoubtedly in part the result of the involvement of modern science with the unimaginably small and the impossibly large. The main impetus, however, is a widespread sense that the world as it is known is a construct. If there is an absolute reality—a *real* reality—it is known only as a tantalizing, ever-receding goal; in Wallace Stevens's words, "It is possible, possible, possible. It must/ Be possible."

The games of modern science are serious and demand total commitment from the players. When John Von Neumann developed his seminal mathematical theory of transactions, a theory with important applications in economics, diplomacy, and national defense, he called it "game theory."

Even though they are serious, however, the games are often so intricate and their rules so strange that the play becomes overtly playful. Hence the fascination of so many scientists with the playful imaginations of children, an interest wonderfully evident in *Mindstorms: Children, Computers, and Powerful Ideas*, by Seymour Papert of MIT. The playfulness spills over into mathematical and logical puzzles and into language that is intentionally paradoxical, whimsical, and absurd. Hence the popularity of playful collections of logical puzzles like *What Is the Name of This Book?* by Raymond Smullyan, philosophy professor and former magician.

Hence, too, the fondness of science for trivial but endearing absurdities. Do you have trouble remembering pi to twenty-two digits? Try the following bit of mathematical-poetical slapstick, in which the number of letters in each word gives you a digit:

> How I wish I could recapture pi.
> Eureka! cried the great inventor.
> Christmas pudding, Christmas pie
> Is at the problem's very center.
>
> [Answer: 3.14159265358979323846]

The archetype and grand model for such logical slapstick is the work of Charles Lutwidge Dodgson, who taught mathematics at Christ Church, Oxford, wrote on non-Euclidean geometries and symbolic logic, experimented with photography, carried on long flirtations with little girls, and wrote poetry and fiction under the name Lewis Carroll.

Carroll's games turn social conventions both figuratively and literally upside down:

> "You are old, Father William," the young man said,
> "And your hair has become very white,
> And yet you incessantly stand on your head—
> Do you think, at your age, it is right?"
>
> "In my youth," Father William replied to his son,
> "I feared it might injure my brain;
> But now that I'm perfectly sure I have none,
> Why I do it again and again."

In *The Annotated Alice* (1970), Martin Gardner suggests that *Alice's Adventures in Wonderland* and *Through the Looking-Glass* are games for grown-up scientists: "It is only because adults—scientists and mathematicians in particular—continue to relish the *Alice* books that they are assured of immortality." If modern scientists keep the *Alice* books alive, they do so because Dodgson's oscillation between symbolic logic and bizarre fictional games anticipates a central theme of modern science. Gardner remarks:

> The last level of metaphor in *Alice* is this: that life, viewed rationally and without illusion, appears to be a nonsense tale told by an idiot mathematician. At the heart of things science finds only a mad, never-ending quadrille of Mock Turtle Waves and Gryphon Particles. For a moment the waves and particles dance in grotesque, inconceivably complex patterns capable of reflecting on their own absurdity. We all live slapstick lives, under an inexplicable sentence of death.

Father William, *by John Tenniel, from* Alice's Adventures in Wonderland *(1865)*.

This is a long way from Thompson's noble numbers, but it underscores nicely the relation between the madcap playfulness of *Through the Looking-Glass* and the playfulness of science.

Had he lived a few more years, D'Arcy Thompson would have been delighted to learn that a long molecule in the shape of a double helix, spiraling upward like the red stripes on an old-time barber's pole, is at the center of life, and that life's infinitely complex forms are spelled out by a code with an alphabet of only four chemical compounds, called bases.

Physics and astronomy refuse to be as neatly geometric as biology. The world they reveal grows more and more like something designed by the Mad Hatter. In its quest for simplifications, particle physics constantly finds new complexities. The particles that are, for the moment, the most fundamental units of matter are called "quarks." Quarks were supposed to simplify things. But the more they have been studied, the more bizarre they have gotten to be. Today they come in pluses and minuses, quarks and antiquarks. There are twelve varieties, which are distinguished by six flavors, three colors, three anticolors, and varying degrees of strangeness. The colors are red, green, and blue, and the anticolors are cyan, magenta, and yellow. One of the flavors has charm; hence the charmion.

The name "quark" was taken by Murray Gell-Mann, the Cal Tech physicist who first proposed quark theory in 1963, from a sentence in James Joyce's *Finnegans Wake*: "Three quarks for Murster Mark." It was an arbitrary and playful act of naming, not unlike the invention of Humpty-Dumpty or the creation of a Dada poem. Quarks are arranged in an eightfold way, a phrase borrowed from Zen. They are combined by particles that are called gluons because they glue quarks together. When gluons stick to other gluons they are called glueballs. Glueballs may (or may not) have been detected in the 1980s at the Tasso Physics Institute in Hamburg. Gluons have color and cause, we are told, either infrared or ultraviolet slavery.

Quarks and gluons have to exist in order for the equations of nature to work out. If they did not exist, nature would be inconceivable or, at least, inconceivably messy. On the other hand, quarks are more elusive, even, than the Cheshire Cat. At the time of this writing, no quark had been observed directly and perhaps none can be.

Yoichiro Nambu, a quark expert at the University of Chicago, remarks mildly, "If quarks are real particles, it seems reasonable that we should be able to see them." The sentence has a nice Baconian ring to it. If quarks are out there, you would like to get at least a glimpse of them as they fly by. Seeing is believing. If you can't see them, you ought, at least, to be able to feel them. That would make them real, like the atoms imagined by Sir Isaac Newton—hard, impenetrable, and perfectly inelastic.

Nambu argues, however, that the theory that proves there are quarks contains within it the corollary that they are forever invisible, or, as he says, "confined." The energy required to dislodge a quark so that it can be observed may change the quark into something else in the very act of dislodging it. Nambu resorts to metaphors to clarify the situation. Quarks are either tied to the ends of a string that cannot be broken or locked up in bags that are impossible to penetrate. This leads to a question that has a true Baconian ring to it: "Theories of quark confinement suggest that all quarks may be permanently inaccessible and invisible. The very success of the quark model leads us back to the question of the reality of quarks. If a particle cannot be isolated or observed, even in theory, how will we ever be able to know that it exists?"

Quarks are now relatively tame inhabitants of the scientific looking-glass world. In the everyday world, people usually assume they know where they are. In other words, they think they understand the idea of location. In 1982 Alain Aspect performed experiments that call the intuitive concept of location into question.

To understand what Aspect seems to have discovered, you need to think again about the particles that Newton so admired. They were not only hard; they were *indivisible*. That means they were little individuals, forever separate, each with its own location, even when they formed committees that would later be called molecules.

Since nothing is supposed to be able to move faster than the speed of light, a birthday card from one particle to another, half a light-year away, can never be delivered in less than six months. That is frustrating if you want to share some juicy gossip, but it is also reassuring. If you could communicate instantaneously with another particle, how would you know you were separate from it? You can say that the idea of location depends on the idea that it takes time to get from one place to another. For example, when the telegraph was introduced, people said it caused the world to "shrink" because it allowed people at different ends of a continent to communicate as if they were in the same room. If you can communicate immediately with someone else without an artificial device like a telegraph, it follows that the two of you must be in more or less the same location.

Aspect's experiments seem to show that certain particles—photons—communicate with one another instantaneously even though they are separated by considerable distances, perhaps even very large distances like light-years. That seems to mean they can be separate but in some way in the same place at the same time. Another Alice-in-Wonderland anomaly.

Is reality continuous, like a chessboard? Maybe it is like a porch screen with more holes than surface or the foam on a glass of beer. Much recent theorizing about the world of subatomic particles has revolved around superstrings. Superstrings are tiny strings, much smaller than a proton, but if they are anywhere, they are everywhere. They are in violent motion. As they move, they create what Michael Green, an expert on the subject, calls a "worldsheet," which is the substratum on which the universe rests or, perhaps, the paper on which it is printed. As superstrings sweep along the worldsheet, they frequently split, leaving holes. Consequently, Green explains, reality is "like . . . a doughnut with an arbitrary number of holes" and spacetime is "foamy."

One of the oddest facts about superstrings is that if they exist they probably have ten dimensions, of which four—the three dimensions of space that you rely on when you walk down to the local tavern plus time—have "uncurled," while six have remained "curled" and unexpressed since the first moments of creation.

If unimaginably small things are odd, the same is true of inconceivably large things. In fact, it is common today to talk of the very big in terms of the very small. This may puzzle laymen, but it is exhilarating to scientists because it suggests a convergence of two apparently very different areas of knowledge—particle physics and cosmology. The goal that shimmers just beyond the horizon of such speculation is what Stephen Hawking, in *A Brief History of Time*, calls "a complete unified theory of everything in the universe."

Much speculation about early cosmology begins with what is called "the big bang," the explosion thought to have created the physical universe. According to principles first worked out in the 1920s by the Russian physicist Alexander Friedmann, it occurred (if it did occur) between ten and twenty billion years ago. That is a long time. What is odd is that to talk about the big bang, you need to talk, not about the passage of billions of years, but about moments so brief that they make a billionth of a second seem like a year on the beach in Bora-Bora.

Exponential notation is a scientific way of writing very large and very small numbers. 10^2 is familiar enough—it is 10 squared, or 1 followed by two zeros: 100. In the same way, 10^5 is 1 followed by five zeroes: 100,000. If there is a minus sign in front of the exponent, the number is a fraction. Thus 10^{-2} is 1/100, and 10^{-5} is 1/100,000. The system is cumbersome for the numbers you need to balance your checkbook, but when the numbers get very big or very small it is a great space-saver, and when you are talking about the big bang, you are talking about something so big it is everything and so small it may, in fact, be nothing.

In 1982 Alan Guth and Paul Steinhardt complained, "Until about five years ago there were few serious attempts to describe the universe during its first second." This oversight is being remedied. Cosmologists have now pushed their theories back to around the 10^{-43} second after the big bang. This probably represents a limit. Something happened before that, but there may be no surviving evidence to show what it was.

Much interest focuses on what is called the "inflationary theory" of creation. This theory, now much modified, was proposed by Guth in the 1970s and elaborated by Steinhardt. It agrees fully with the big bang theory about what happened *after* the first 10^{-30} second, but what happened before that makes all the difference. It assumes that immediately after the big bang, a transition occurred during which the universe took shape more or less as a hole forms in a Swiss cheese or a bubble in a glass of ginger ale. Our universe is one of many bubble universes which may "continue to form forever." For a time, the bubble expanded rapidly. It is this period of "inflation"—rapid growth—that gives the theory its name. After the inflationary period another transition occurred. Matter precipitated out of the vacuum inside the bubble, and the universe was irreversibly on the way to becoming its familiar self.

One drawback of the inflationary theory is that if it is true, no consequences of anything existing before the inflation can ever be observed. They are forever beyond our horizon of invisibility. The problem is similar to that raised by Yoichiro Nambu about quarks: "If a particle cannot be isolated or observed, even in theory, how will we ever be able to know it exists?"

On the other hand, the inflationary model has a splendid virtue. Because it treats matter as a positive form of energy and gravity as a negative form of energy, everything that exists can be canceled out by something else that exists. The sum is breathtakingly elegant but in a way that might have made D'Arcy Thompson unhappy: zero. Having quoted with approval a suggestion that the universe originated "as a quantum fluctuation from absolutely nothing," Guth and Steinhardt add in their 1982 article that it is "tempting to go one step further and speculate that the entire universe evolved from literally nothing." Guth has gone the extra step. "It is said," he once remarked, "that there is no such thing as a free lunch. But the universe is the ultimate free lunch."

Quantum theory shows that nature cannot be separated from the person observing it. Quark theory suggests the existence of entities that can never be observed. By proposing that everything in the universe comes from nothing, the inflationary theory makes the disappearance of nature official. We are such things as dreams are made of, but who is doing the dreaming? "Now, Kitty," said Alice, "let's consider who it was that dreamed it all. This is a serious question, my dear, and you should *not* go on licking your paws like that."

6

MANDELBROT'S MONSTROSITIES

To many people, the world of quarks and black holes is an affront. Humanity seems to have leaked out of it. To others it is a playful world—a world of games and of necessary fictions, though some of the fictions are more necessary than others. The choice of fictions may be decided by a throw of the dice, but perhaps not. The question is probably being asked in terms of a past that no longer exists rather than a present that may not exist but is as good a bet as any when you are playing against the house.

Karl Friedrich Gauss and Nikolai Lobachevsky, both brilliant mathematicians, began investigating non-Euclidean geometries early in the nineteenth century, and Georg Friedrich Riemann placed the enterprise firmly on its modern course. The results stirred bitter controversy within as well as outside of the mathematical fraternity because they challenged the ideal of an eternal order of things beyond nature. In effect, they made geometry into a game.

Geometry and play are wonderfully interwoven in Benoit Mandelbrot's *Fractal Geometry of Nature* (second ed., 1982). Like D'Arcy Thompson, whom he cites, Mandelbrot writes for the general reader as well as the specialist. Although his ideas are complex, he presents them in nontechnical as well as technical form. By inviting the reader to skip sections that are "outside his interest or beyond his competence," he makes a potentially forbidding subject inviting and accessible. The informality is a little like the invitation of the barker at a fun house. As you come to understand it, Mandelbrot's mathematics is closer to the surrealistic fantasies of André Breton or René Magritte than D'Arcy Thompson's stately Pythagorean forms.

Mandelbrot's central concept, which gives his book its title, is that the notion of three simple dimensions is a myth. Real-world objects occupy a space whose dimensions are fractional, or, as Mandelbrot puts it, "fractal." One kind of "less-than-three-dimensional" object may have 1.2618 dimensions, more or less; another may have 1.6131, more or less. Dimensions of one, two, and three are theoretically possible but abnormal.

Bacon and Newton would have been as shocked as Kepler and D'Arcy Thompson by this notion. It is worse than the idea that things can be called real that even in theory are anything but hard, impenetrable, and massy; and that cannot be observed anyway; and that, if observed, would probably turn out to be made out of nothingness. If the idea of fractals is shocking in three dimensions, what might it be in four or in the ten-dimensional space of superstrings? No wonder that traditionalists called the sort of mathematics found in *The Fractal Geometry of Nature* "pathological" and "a gallery of monsters." It is a measure of how far nature has gotten out of control in the last quarter of the twentieth century that precisely such pathological and monstrous devices should be offered without irony as the best way of representing it.

The aesthetic appeal of science is evident in the admiration of Darwin, the Baconian, for the "beautiful co-adaptations" of birds, in the awed contemplation of divine geometry by Pythagoreans like Kepler and Thompson, and in the simple sense of the beauty of order expressed in the writings of scientists like Newton and Einstein. Darwin's ambivalence about his own powerful aesthetic responses to nature is one of his striking weaknesses and is directly related to his blurred sense of where the observer stops and nature begins, as evidenced in his constant projection of human motives into nature. Thompson seems, on the surface, more consistent. He recognizes that the mathematics he considers the ground of the real is a product either of man or of God, with the odds at least officially on God. He is therefore able to accept the aesthetic element in science more easily than Darwin.

Mandelbrot is deeply concerned with the aesthetics of science. His fractal geometry suggests that Thompson was as confused as Darwin though in a different way. Thompson's elegant and regular geometries are not so much in nature as imposed on it. They are fictions and not even necessary ones. Mandelbrot argues that those who use geometric fictions to interpret nature are actually rejecting it both scientifically and aesthetically. They have "increasingly chosen to flee from nature by devising theories unrelated to anything we can see or feel."

For Mandelbrot, "Euclid" is therefore a bad word, synonymous with a dead, unnatural formalism. "Why," he asks, "is geometry often described as 'cold' and 'dry'? One reason lies in its inability to describe the shape of a cloud, a mountain, a coastline, a tree. Clouds are not spheres, mountains are not cones, coastlines are not circles, and bark is not smooth, nor does lightning travel in a straight line."

As we have observed, traditional geometers, confronted by shapes that refused to behave in the well-brought-up manner of Euclid's circles and parabolas, called them "terrifying" and "a gallery of monsters." Mandelbrot argues that the likeness to nature of his fractal shapes makes them more, not less, beautiful than Euclid's regular forms. Freeman Dyson, a reviewer of *Fractal Geometry*, relates the rise of interest in unorthodox mathematics to twentieth-century art when he compares the impulse to study to the impulse behind "Cubist paintings and atonal music that were upsetting established standards of taste in art at about the same time." Perhaps the impulse was similar, but Mandelbrot rejects Cubism along with Bauhaus architecture: "A Mies van der Rohe building is a scalebound throwback to Euclid."

The appeal to "likeness to nature" does not imply that Mandelbrotian mathematics will resurrect the collection of things in the middle distance that was the object of Baconian science. Quite the contrary, Mandelbrot is not out to reproduce nature but to make convincing models of it. Verisimilitude is part of his demonstration of the rightness of his approach. Inherent in his mathematics is the assumption that nature is infinitely complicated and variable. It can never be "observed." It can only be approximated.

Brownian movement is the zigzag movement of tiny particles like dust motes as they are buffeted by collisions with molecules of a gas or a liquid. It is out there in nature and apparently available to observation. You can watch those little motes zigzagging around before your eyes. Can you really? What you see during any observation are two positions of your mote—a "before" and an "after." Say it goes from point "A" to point "B." If you make more frequent observations of your mote, you will discover that in going from point "A" to point "B," it goes from point "A" to point "Z" and then to point "Y" and *then* to point "B." What is curious and a little frightening is that every time you make your observation more precise, you discover new zigzags—and the increasing complexity of the movement seems to have no limit. Mandelbrot explains: "One may be tempted to define an 'average velocity' . . . by following a particle as accurately as possible. But such evaluations are *grossly wrong.* The apparent average velocity varies crazily in magnitude and direction. . . . If [the] particle's position were marked down 100 times more frequently, each interval would be replaced by a polygon smaller than the whole [previous] drawing, but just as complicated, and so on."

In other words, if you draw the movements of a particle at one scale and then make a drawing of a much magnified part of the first drawing, the second drawing will be just as complicated and just as full of jerky jumps as the first. The same thing will be true for a drawing of the drawing of the drawing, and so forth. You apparently never get to the point where you can say, "This is it; after this, all movements are regular and predictable."

Brownian movement is a good symbol for the view of nature suggested by Mandelbrot's geometry. An absolute barrier stands between man and nature. Mandelbrot does not offer clever ways of climbing over the barrier, only new and better ways of producing imitations of what is on the other side. This is another way of saying that nature *per se* is invisible. We have to be satisfied with facsimiles, and facsimiles are human artifacts.

In one respect, Mandelbrot is taking a position not unlike the one forced on quantum mechanics in the wake of Heisenberg's uncertainty principle. Mandelbrot is aware of this. He remarks, "The notion that a numerical result should depend on the relation of object to observer is in the spirit of physics of this century and is even an exemplary illustration of it." Reality is partly our own creation, and what we see in it is partly the shape of our own motives. To admire the fractal images in Mandelbrot's book is to admire creations that admit their human origin as frankly as a painting by Matisse.

To get an idea of what a fractal is, consider a line that squiggles around on a page—a doodle you might make, for example, while talking on the phone. As described by traditional geometry the squiggly line has one dimension—length. The surface on which it is written has two dimensions, length and width, and the space you occupy while doodling has three dimensions. The conventional dimension of a point is "0"; of a line, "1"; of a plane, "2"; and of the space around your telephone table, "3." "1," "2," and "3" are integers in contrast to a number like "1.5," which is a fraction.

If the phone call lasts a long time, your doodle will get more and more complicated, and eventually it will almost cover the page. If the phone conversation lasted forever (which never happens except among teenagers), it *would* cover the page, and its dimension would change from "1" to "2." In order to describe the tendency of a squiggly line to become a surface, Mandelbrot says that it has a *fractal dimension.* A fractal dimension is also called a "Hausdorff dimension" after Felix Hausdorff, who developed the idea in 1919. The fractal dimension for the doodle might be something like 1.5—sort of a halfway point between "1" and "2."

Lines with dimensions of 1.5 probably seem as bizarre as you can get. But are they? We like to pretend we live in a tidy world full of circles and triangles and right angles. Anybody who has tried to make a picture frame knows that right angles are the exception rather than the rule in the real world. Mandelbrot calls the world of lines and planes and right angles "Euclidean" because Euclid is constantly using it in his geometry. Mandelbrot considers it a fantasy, and an unnatural one at that: "Many patterns of Nature are so irregular and fragmented that compared with

Euclid—a term used in this work to denote all of standard geometry—Nature exhibits not simply a higher degree but an altogether different level of complexity."

Take coastlines. Mandelbrot's fifth chapter introduces the subject with the question "How long is the coast of Britain?" A good question, which, as Mandelbrot notes, was first asked by Lewis Richardson, mathematician, advocate of world peace, and uncle of Sir Ralph Richardson, the Shakespearean actor.

If you look at a map made at a scale of one hundred miles to an inch, the coast is obviously not smooth. It goes in and out in bays and promontories and estuaries and capes. You include these when you measure it. If you use a map drawn at a scale of ten miles to an inch, new bays suddenly open up in the coastlines of promontories, and new promontories jut out from the sides of bays. When you measure these and add them to your first total, the coast gets longer. It gets longer still at a mile to an inch—and so on until you are crawling around on your hands and knees measuring the distances around small rocks. If you decide to use a microscope, you will find yourself measuring the irregularities on the surface of each rock and . . .

Enough! Mandelbrot has been faithful to nature, but where has the coast of Britain gone? Is there a coast? Is the problem serious or absurd? It is certainly playful, like the logical puzzles in *Through the Looking-Glass*. It also has practical implications. Mandelbrot compares the length of the border between Spain and Portugal in different atlases. The Portuguese atlas shows the border as 20 percent longer than the Spanish atlas. Should Spain break off diplomatic relations with Portugal? No. Both atlases are correct. The Spanish surveyors based their measurements on a larger unit of distance than the Portuguese and therefore measured fewer squiggles in the line defining the border.

To analyze the coastline of Britain (or the border between Spain and Portugal) mathematically, Mandelbrot creates a fractal line that behaves like a coastline. In other words, he makes a mathematical model. He begins with a regular shape—an equilateral triangle. He then introduces a regular deformity—an equilateral triangle—into each of its three sides. The result is a twelve-sided figure shaped like the Star of David. Next he repeats the operation on each of the twelve sides, creating a figure

Benoit Mandelbrot, successive forms of the Koch Triangle *(1987).*

with forty-eight sides, and so on until the changes are so small the eye cannot follow them.

Since the process of adding triangles is repeated each time the scale is changed, the result is a series of figures that have "self-similarity." The triangle never stays still because it has no bottom. Every time you try to measure it you discover that at the next step down it has squiggles you left out in your last measurement. It is not a coastline, but it is like a coastline. It is called a Koch triangle, and on a scale of one to three dimensions, it has a fractal dimension of about 1.26.

Clouds, riverbanks, coastlines, tree branches, the branchings of ever-smaller air passages in the lungs, erosion systems, commodity prices, tree bark, word frequencies, turbulence in fluids, stars in the sky, and galaxy clusters in deep space are all wondrously, dizzily, and irreducibly fractal.

Or, to put the idea more correctly, all *seem* to be fractal. Something very interesting and wonderful happened while Mandelbrot was measuring the coastline of Britain. He was not observing nature but devising ways to use mathematics to generate things "like" nature. His shapes are necessary fictions. Their claim to truth value is based primarily on their fidelity to the formulas that created them. In a curious secondary way, their claim is also based on their similarity to the bits of nature they represent.

Artists and poets have been urged to imitate nature ever since the Greeks. One test of Mandelbrot's models is "likeness" to their originals. The test is often disarmingly direct. Seeing is believing. Does the picture look like the thing pictured? Although randomness is not essential to fractal mathematics, random irregularities are a common feature of natural phenomena. Mandelbrot regularly introduces random irregularities into his constructions. The irregularities make them more "like" the clouds and mountains that they imitate and less like triangles and spheres.

When randomness is part of the generating process, Mandelbrot's squiggles are no longer entirely predictable. The squiggles at one level of magnification are different from those at another, although at all levels the squiggles may have a family resemblance. The variations make them more rather than less like nature. When the irregular squiggles enclose areas, they look uncannily like coastlines of islands or continents or shorelines of lakes. When they are three-dimensional, they look like natural landscapes. The resemblance is so striking that fractal geometry is routinely used to produce landscapes of unknown worlds by studios like Lucasfilm. The likeness carries over into phenomena like eddy currents and turbulence. Mandelbrot would argue that the presence of so much likeness is a strong indication that randomness is part of the deep structure of nature. Again, "Seeing is believing." (For an illustration, see p. 222.)

To create a fractal shape you have to add squiggles to squiggles to squiggles. In other words, you have to perform the same operations over and over again, just as the irregularities of a Koch triangle are repeated with each new cycle until the computer is turned off. In theory the operation can go on forever. The squiggles can be enlarged repeatedly until the line covers the sky. By the same token, they can be diminished repeatedly until they resemble strands of DNA. There is obviously a limitation. The generation of complicated squiggles by hand quickly becomes impossibly laborious. For this reason there is a close symbiosis between computers and the development of fractal geometry. Since computers never get bored repeating themselves and will plot complicated functions in color and with wonderful accuracy, all of the more striking fractal images, whether of landscapes or of abstract forms, are created by computer graphics programs.

The result can go endlessly upward toward an infinity of the large and endlessly downward toward an infinity of the small. As the example of Brownian movement has shown, there are apparently many such situations in nature. You can regard them with amusement or awe. Mandelbrot does both. Amusement is evident in his quotation from Jonathan Swift's "On Poetry: A Rhapsody" (1733):

> So Nat'ralists observe, a Flea
> Hath smaller Fleas that on him prey,
> And these have smaller Fleas to bite 'em,
> And so proceed ad infinitum.

Awe is suggested by a quotation from Immanuel Kant's *Universal Natural History and Theory of the Heavens* (1755):

> It is natural . . . to regard the [nebulous] stars as being . . . systems of many stars. . . . [They] are just universes and, so to speak, Milky Ways. . . . It might further be conjectured that these higher universes are not without relation to one another, and that by this mutual relationship they constitute again a still more immense system . . . which, perhaps . . . is yet again but one member in a new combination. We see the first members of a progressive relationship of worlds and systems; and the first part of this infinite progression enables us already to recognize what must be conjectured of the whole. There is no end but an abyss . . . without bound.

Mandelbrot recognizes that in nature there are often limits that prevent the process from moving toward the infinites of the large and the small. He calls them "cutoffs." They resemble phase transitions in physics. Mandelbrot illustrates cutoffs by describing an imaginary observer's view of a ball of thread. Only once in the series of observations is the observer in the middle distance where the ball of thread is what it is supposed to be:

> A ball of 10 cm diameter made of a thick thread of 1 mm diameter possesses (in latent fashion) several distinct effective dimensions. To an observer placed far away, the ball appears as a zero-dimensional figure: a point. . . . As seen from a distance of 10 cm resolution, the ball of thread is a three-dimensional figure. At 10 mm, it is a mess of one-dimensional threads. At 0.1 mm, each thread becomes a column and the whole becomes a three-dimensional figure again. At .01 mm, each column dissolves into fibers, and the ball becomes one-dimensional, and so on, with the dimension crossing over repeatedly

> from one value to another. When the ball is represented by an infinite number of atomlike points, it becomes zero-dimensional [because a point is said to have 0 dimensions].

The observing of the ball of thread is segmented by the cutoffs. Because of the transitions, nature as found in the ball of thread is "grainy"—different systems appear at different scales. Part of the difference between one phase and the next is objective—the observer is always seeing what is there. But part is subjective—the observer chooses the scale, and in this sense he chooses what he will see. Again we are reminded of the relation between observer and observed in fractal mathematics.

How long is the coast of Britain? What are you after? Are you a space shuttle or a cruise ship captain or a fisherman in a rowboat? Mandelbrot remarks, "In one manner or another, the concept of geographic length is not as inoffensive as it seems. It is not entirely 'objective.' The observer inevitably intervenes in its definition."

In Heisenberg, indeterminacy is decently submerged in the spaces between (or among) electrons. In Mandelbrot, it is swimming boldly along on the surface like some Loch Ness monster or self-squared dragon circumnavigating the coast of Britain.

Fractals are obviously mathematics. Are they science in any other sense of the term? The answer is that they have an astonishingly broad series of applications, from measuring stock market performance to weather forecasting to the analysis of contact between surfaces like tires to road, to fracturing of glass and metal. The branchings of veins in the surface of a leaf, of air pipes (alveoli) in the lungs, of neurons in the brain are fractal. E. I. Du Pont discovered how to manufacture good-quality synthetic goose down for sleeping bags when the company chemists realized that natural goose down was fractal in structure. There are reasons to believe that evolution may be a fractal process with phase transitions in the form of mutations.

In *Chaos: Making a New Science* (1987), James Gleick outlines the branch of mathematics called "chaos theory." Chaos theory draws heavily on fractal mathematics, a good example being the analysis of the onset of turbulence in water flow.

The fractal world is so strange that Mandelbrot has devised a new poetry to name its citizens. He obviously enjoys the task. It is another

Leaf showing fractal branching of veins. Detail of photograph by Kjell Sandved, from Leaves *(1985).*

kind of game. The word *fractal* is part of his poetry. It is derived from the Latin *frangere*, "to break," "to create irregular fragments." Other coinages are dust, curd, and whey; grainy structures, hydralike structures, ramified structures, tiled surfaces, and pertiled surfaces (in which collections of little tiles make big tiles like the little tiles of which they are made); pimply, pocky, wrinkled, and wispy surfaces; self-squared dragons, monkeys' trees, Minkowski sausages, flattened flowers, skewed webs, random slices of Swiss cheese, and chains and squigs.

Benoit Mandelbrot, image generated using Mandelbrot equations.

One of Mandelbrot's most striking constructions is pure surrealism. It is a "self-squared dragon," and it is the ancestor of a large and exotic line of computer monsters, some in four rather than three dimensions (see illustration 3 in color section). Alan Norton, a colleague of Mandelbrot at IBM's Thomas J. Watson Research Center, describes photographing the four-dimensional dragons that float up from his computer screen as "throwing my camera out there in the dark, taking snapshots."

The ultimate proof of any pertiled pudding is its eating. *The Fractal Geometry of Nature* offers picture after picture of breathtaking playfulness and astonishing likeness—sometimes to itself and sometimes to nature.

Is it art? The question is significant. Is Kepler's image of the planetary orbits in relation to four regular solids art? Is Darwin's description of the La Plata woodpecker art? Are Audubon's bird paintings art?

Mandelbrot thinks it is art—a "new geometric art." Pictures are essential to Mandelbrot's text. They are often strange, but even the most bizarre of them teases the mind with its familiarity. "In the theory of fractals," writes Mandelbrot, " 'to see *is* to believe.' . . . The reader . . . is . . . advised to browse through my picture book." If Darwin is an artist in words and D'Arcy Thompson an artist of regular forms, Mandelbrot is an artist of the games nature plays and the masks it wears.

He obviously feels that fractal geometry has the potential of becoming a full-fledged art form. True to his opposition to Euclid, he contrasts the richness of fractal art with the "minimal art" of "lines, circles, spirals, and the like." Fractal art is antiminimalist and thus akin to "Grand Masters paintings or Beaux Arts architecture." To illustrate the superiority of fractals, he cites traditional paintings that embody fractal techniques along with mathematical constructions: God the Creator, from a twelfth-century Bible manuscript; the waters of Noah's flood, by Leonardo da Vinci; and *The Great Wave*, by Katsushika Hokusai (1740–1849). Mandelbrot concludes, "For all these reasons, and also because it came in through an effort to imitate Nature in order to guess its laws, it may well be that fractal art is readily accepted because it is not truly unfamiliar."

Fractal mathematics was born from an effort "to imitate Nature and guess its laws." Has not imitation of nature—however nature may be understood—been the special duty of the artist since the Greeks? At the same time, nature, as far as Mandelbrot can represent it, is radically elusive and probably monstrous and without doubt terrifying to any right-

Satellite photograph with fractal erosion patterns (NASA).

minded disciple of Euclid. Does nature exist? As far as fractal geometry is concerned we can never know and we do not need to bother.

The science of the late twentieth century asks man to understand himself in the light of his own reason detached from history, geography, and nature, and also from myth, religion, tradition, the idols of the tribe, and the dogmas of the fathers. It offers likenesses of nature, not nature, and it suggests further that nature is a project created in part by man. Culture is an artifact and probably a game, and what happens in it is the result of human rather than divine will.

Objectifying this understanding of things requires new languages. Mallarmé, perhaps the greatest French poet of the nineteenth century, understood this. The speaker in his masterpiece "Un Coup de dés"—"A Throw of the Dice"—creates meaning by random gestures. If reality is a game, Mallarmé's player is more like a gambler at 3:00 A.M. with a stack of IOUs on the table than a child with a new toy, but only a few years later, in the art of the early twentieth century, the players were having fun and reveling in the newfound freedom to change the rules and invent new games as they went along. A similar spirit in modern science is reflected in the naming of quarks and superstrings and self-squared dragons. The names are playful because play is an essential part of the activity that gives rise to them.

If science is a human creation, we have caught the mind in the very act of swallowing up the world, which is another way of saying that we have witnessed nature in the process of disappearing. The steps are neatly defined by the figures of Charles Darwin, D'Arcy Thompson, and Benoit Mandelbrot. They take us from a nature that is alien and into which human motives are poured, to a nature that is number—but number authenticated by an absolute order—to an imitation of nature by means of number that is also a form of art. Darwin's world is mythic. Mandelbrot's is pure fiction, which means that it is entirely human. Games are human inventions. A throw of the dice will never eliminate chance, but it keeps the games interesting.

PART II

GREAT WALLS AND RUNNING FENCES; OR, THE DISAPPEARANCE OF HISTORY

The nasal whine of power whips a new universe . . .
Where spouting pillars spoor the evening sky,
Under the looming stacks of the gigantic power house.

—HART CRANE,
The Bridge

Georges Vantongerloo, "Composition Derived from $y + ax^2 + bx + 18$*. . . No. 62" (1930).*

7

HISTORY IN WASHINGTON, D.C.

The most impressive structures surviving from antiquity are temples, monuments, and palaces. The ziggurats of Assyria, the temples of Luxor, the Athenian Parthenon, the Pantheon in Rome, the pyramids of Yucatán, and the temple complex at Angkor Wat have their counterparts in Chartres and Notre Dame in France, St. Sophia in Constantinople, St. Peter's in Rome, and St. Paul's in London. They are the homes of the creating and sustaining gods and stages for the rituals through which men communicate with those gods.

The political order also has its monumental structures. They are the tombs, palaces, and government buildings that celebrate dynastic and national power: the pyramids of Egypt, the palace complex in Beijing, the Roman Forum, the Kremlin, Versailles, the houses of Parliament in London, the Capitol Building in Washington, D.C.

Monumental architecture grows out of a native soil. It expresses religious and historical traditions specific to the culture in which it appears. To take the example nearest to Americans, the dominant style of the government architecture in the District of Columbia is classical because its builders wanted to say something about the historical roots of America's government.

When Major Pierre-Charles L'Enfant was designing Washington, he concluded that there were exactly seven hills in the territory to be occupied by the new city—just the number that ornamented ancient Rome. By happy coincidence, one of these hills was at the center of the future city and drained by a small stream, long since covered over, called Tiber Creek. When L'Enfant considered this hill he knew immediately what it was for. It was "a pedestal waiting for a monument." Its name was accordingly changed from Jenkins Hill to Capitol Hill, after the Capitoline Hill in Rome.

The monument resting on the pedestal is the Capitol Building. Its classical style announces from afar that it is ground zero of the metaphysical space of a great nation. The fact that Washington's streets radiate outward—north, east, south, and west—beginning with A, C, and D (B is mysteriously left out) and 1, 2, and 3 makes the alphabet and the real number system bear witness to this truth. The dome that surmounts it and later, as the perspective sharpens, the rows of classical columns, surmounted by architrave and pediment, that appear on every side are another statement of the same theme.

They assert that the underlying religion, as well as the political system, of America is not an invention of the sort that might occur to the naïve on the basis of the phrase "New World." America's civic religion and its political order are linked by the Capitol Building to the system that the Founding Fathers derived—or thought they had derived—from the republican traditions of Greece and Rome.

A dome is a specifically Roman form. Domes were often used for temples, and the best preserved and most copied dome surviving from antiquity is the one that covers the Pantheon—"the home of all the gods"—in Rome. Michelangelo adapted this concept when he designed the dome that at present crowns St. Peter's Basilica in Rome. His design

Dome of the Capitol, Washington, D.C.

became, in turn, the inspiration for the dome of the Panthéon in Paris and the cast-iron dome of St. Isaac's Cathedral in Leningrad.

In 1851 the American Congress decided to replace the drab, saucerlike dome designed for the Capitol by William Thornton with a larger, more imposing one. The second dome is the work of Thomas Walter. Its form recalls the form of Michelangelo's dome for St. Peter's. When the Civil War broke out, proposals were made to halt its construction. President Lincoln insisted that construction continue because, he said, "If the people see the Capitol going on, it is a sign we intend the Union shall go on."

That in itself invests the dome with the qualities of a temple. It symbolizes the preservation of the Union. Domes are also, in their original symbolism, representations of the sphere of the sky and thus of the boundary separating visible from invisible reality. Domes also represent mortality because the visible world is subject to time, while the invisible world is timeless. The outside of the last of the spheres in Fludd's illustration of the Ptolemaic universe is populated by angels. (See p. 8.)

A small circular opening is usually found in the center of an ancient dome—the "oculum," or "eye." The opening allows spirits to travel to and from invisible reality. The interior of a dome is thus a womb and the opening a cervix through which the soul can be born into eternity. Often the oculum is protected by a circular structure called a lantern and surmounted by the statue of a god. The Capitol dome has no oculum, but it has a lantern, surmounted by Thomas Crawford's heroic statue of *Freedom* clasping a sword and shield. Quite proper. *Freedom* is presiding goddess of America's civic religion.

The interior of the Capitol dome is covered by a 4,664-square-foot painting by Constantino Brumidi (1805–1880) entitled *The Apotheosis of George Washington*. In this painting the soul of Washington is shown rising toward what should be the eye of the dome. Around the base of the picture, representatives of the arts and sciences bid Washington a loving farewell—Benjamin Franklin, Thomas Fulton, Samuel F. B. Morse. Washington is in mid-flight, just before he passes upward to join the goddess Freedom in the world beyond time. He is flanked by Liberty and Victory, and around him, their hands linked, are thirteen loosely robed maidens representing the thirteen original colonies.

Statue of Freedom *on Capitol Dome, Washington, D.C.*

When congressional office buildings were built to the north and south of the Capitol, what could be more natural than to continue the classical program? In fact, classical decor was used as late as the Sam Rayburn House Office Building, completed in 1981. Long before then, however, history had begun to lose its power over the American imagination.

The loss announces itself gradually at first and then abruptly. As you walk east along Constitution Avenue opposite the Capitol, you encounter the first of three Senate office buildings, the Russell Building. Completed in 1933, it is resolutely and elegantly classical, with rounded and fluted columns, rusticated lower walls, bowed windows, and elegant porticoes defining its west and east entrances. Next comes the Dirksen Building, completed in 1958. It, too, is classical, but its details are generalized. The walls are flat. The windows are square and defined by panels of dark metal. Strips of flat marble separate them, weakly claiming to be a row of columns. The Dirksen Building is not a tribute to history. It is a tribute to thrift. It is also an anachronism. To move from the Russell Building to the Dirksen Building is to watch a history that is disappearing but has not entirely vanished.

The Dirksen Building stops abruptly in mid-block. For years its eastern end was a flat, entirely plain surface. The building looked like a loaf of bread cut in half. The assumption conveyed by this sudden termination was that a second half of the loaf would eventually be added so that the resulting building would cover the whole block.

It did not happen. When an extension was planned in the early 1970s, the planners decided to throw history away.

Instead of finishing the Dirksen design by respecting its roof line and the strips of marble that claim to be a colonnade, the planners created an entirely new structure opposed to the old one in scale, building lines, and decor—the Hart Building.

The abruptness of the termination of the Dirksen Building makes its rape by the new building all the more obvious.

There is not a single column, architrave, pediment, tympanum, triangular or rounded window header, statue, or frieze anywhere on the façade of the Hart Building. Instead, row on row of two-story rectangular marble boxes parade on the façade as sunscreens for the windows behind them. The use of marble can be interpreted as an allusion to tradition, but the boxes seem to be megalomaniac exaggerations of similar boxy

Junction of Dirksen with Hart Senate Office Building, Washingon, D.C. Hart Building designed by J. Carl Warnecke (1974). Photograph by Stavros Moschopoulos/Image Ray.

enclosures on the façades of late International Style motels. As for roof line, the Hart Building is nine stories high and towers above the building that it so ungraciously joins. The contrast makes the Dirksen Building look squat and mean. At the same time, the Dirksen Building makes the Hart Building look boxy.

When Ada Louise Huxtable, architectural critic for *The New York Times*, reviewed the plans on June 24, 1974, she imagined the architecture speaking. "Look, boys," she wrote, "no hands (columns), but we are using proportions and forms that with a little double-talk and double-take create the illusionist trick of classical order. At the same time, we are being true-to-ourselves modern." Huxtable answered the architecture: "No way. The result is an aesthetic bastard by any measure."

The interior of the Hart Building is equally defiant. Marble is used in large, featureless, highly polished slabs. There is an atrium. Is this a concession to history—an allusion to the enclosed spaces at the center of ancient Roman villas? Perhaps. But the Hart Building's atrium is square and formal. It is not a sanctuary but a void. A menacing abstract sculpture rises from its floor, a Calder made of black steel slabs—*Mountain and Clouds*. Even though the atrium is huge, the sculpture manages to be too big for it. What seemed cavernous without it seems cramped and inadequate now that it is in place.

The Hart Building is featureless, cheerless, polished, cavernous, and crammed at the same time. In spite of these inadequacies, it makes a powerful and explicit statement. It says things have changed. Even the politicians in Washington have gotten the word. It is a word you hear in many other areas of modern culture: history is bunk.

8

THE GREAT WALL SYNDROME

The contrast between the Dirksen Building and the Hart Building is gross, palpable, and unignorable. It dramatizes a central feature of the modern aesthetic, but it does so in an atypical way. More typical is the situation in which the modern aesthetic insinuates itself without forcing awareness of its arrival. For reasons outlined below, this can be called the Great Wall Syndrome.

Technological culture is constantly introducing useful materials and objects into the world. Because they are useful, they are accepted without thought. Sometimes they are so retiring that they are hardly noticed unless they are being used. Their relative invisibility has very little relation, however, to their influence on the shape of consciousness. A telephone is the tiniest tip of an iceberg projecting into a hallway. It is small compared even to the vases on the tables and the pictures on the walls. Who notices a telephone except when making a call? However, the cultural implications of the telephone are enormous. Its wires are connections to a network that covers the planet.

Because it is bigger than a telephone, an automobile seems to make more difference. It is shiny and expensive. Also, it demands roads and eventually dual-lane highways and massive architectural works like cloverleafs and overpasses. We know automobiles have insinuated themselves not only into our garages but also into our consciousness. So have all of the other materials and objects and procedures introduced into modern culture by technology.

Yet as long as they are useful, the Great Wall Syndrome makes them hard to see.

Imagine that when the Great Wall of China was being built by Shih Huang-ti, it was simply a great wall. It spoke only two words: "Keep out." As everyone knows, eventually, it ceased to have a military use. There were no enemies—at least in the north—to keep out. But it was still massively there. Since it could no longer be understood as a wall—that is, as something that keeps somebody out—it had to be looked at. What, after all, was it there for? When this question was asked, those who looked saw it was a wonder of the world. The wonder had been there from the beginning and must have touched all those who passed by, but as long as it was concealed under the veil of wallness, there was no word for the wonder. It was a silence.

The head-on collision of the Hart Building with the Dirksen Building makes the statement of the Hart Building impossible to ignore. But the most pervasive assertions of modern aesthetic are silent. They are the expressions of myriad objects that—like telephones—are humble and apparently unassuming and almost invisible and hide themselves under the veil of utility. Such objects only reveal their natures when they are named and thus forced to stand and identify themselves—as, for example, the Great Wall only reveals its identity when it is called a Great Wall rather than a wall. Understanding the aesthetics of modern culture requires a constant willingness to search for Great Walls under various kinds of wallness.

The Hart Building dramatizes the turning away of modern culture from history. It does so in a crude and belligerent way, as though it could only make its point by belittling the values of earlier architecture, but crudity does not make a statement a lie. The Hart Building repeats a truth told with more sophistication by other and better buildings. The same truth is asserted by the myriad utilitarian products of technical culture. Corrugated steel has no classical precedent, but it is just the thing for cow sheds and storehouses. A bathroom is a bathroom and a grain elevator is a grain elevator. No need for domes or columns or allusions to the Areopagus of Athens or the Senate House of Rome.

To reject history is to move in the direction of abstraction, and a kind of abstraction was thus inherent in the objects produced by technology long before an aesthetic of the abstract had been recognized. Economics reinforced this bias. Rational planning required cost-effectiveness, and historical symbolism was not cost-effective for most utilitarian objects.

Although technology is inherently abstract, the abstraction it produces is, for the most part, as hidden by utility as the Great Wall under its wallness, until art gives it a name. Once it is named, art—including architecture—begins to ask questions: Who needs historical symbolism? Why pay for frills that are at best useless and probably ridiculous to boot? Let things be what they are. Does not the old symbolism, in fact, perpetuate myths associated with archaic superstitions and discredited political systems—with patriarchal oppression, sexism, class tyranny, colonialism, racism, and so forth? Even if it is harmless, is it not extravagant and thus wasteful?

There is an alternative that may seem, on first consideration, a compromise with history. Once the old symbols are no longer regarded as having sacred content or—what is much the same thing—as being true, they no longer need to be considered threatening and can be regarded as attractive curiosities of the sort you might come across in the shop of an art dealer who sells stained glass from Turkish churches along with native crafts of Sicily and rhinoceros horns and pre-Columbian antiquities. In this mode, the old symbols become design motifs and conversation pieces and *curiosa* to be displayed individually or combined into visually attractive mosaics and collages.

Such collocations are often playful. They comment ironically on the draining of historical meaning from the symbols used. A Navaho sand painting, for example, may be hung between a reproduction of a Russian icon and a nineteenth-century Japanese print of concubines entertaining a customer. A table lamp in the shape of a Doric column sheds its rays on a piece of Chinese embroidery using the forbidden stitch, which, in turn, covers a pillow on a restored Morris chair. Detached from their contexts, symbols are liberated from history as completely as the abstract objects produced by functionalism. They become designs.

The prophet of futurism, F. T. Marinetti, vows in his *First Manifesto* to free Italy, and later the world " . . . of its smelly gangrene of professors, archaeologists . . . and antiquarians. . . . We mean to free her from the numberless museums that cover her like so many graveyards." It is well said. The theme of liberation from history links modern art with the technological aesthetic and with all of those presences created by technology that silently but powerfully shape consciousness in ways unnoticed in traditional attempts to understand culture.

Marinetti is celebrating the discovery of an amazing and delightful profusion of Great Walls where others had seen only a collection of drab utilitarian shapes—as one prominent critic put it, "an Arizona of the mind." Maybe Arizona has more delights than it has been given credit for. Maybe history *is* bunk. The pros and cons of this suggestion are explored by many forms of art, but nowhere has the exploration been more self-conscious and more insistent than in architecture, which—as the architecture of the Capital shows—has traditionally had a special relation to history.

9

PARIS DISHONORED

The Crystal Palace is often called the first example of truly modern architecture. It was built in 1851 to honor the triumphs of Victorian civilization, by which was meant Victorian technology. Like the Philadelphia Centennial Exposition of 1876, the Great Exhibition was intended to show off the machines that were the basis of England's industrial prowess. The building itself was intended to demonstrate new applications of technology and it succeeded admirably. In an article in the October 1984 *Scientific American* on its design features, Folke Kihlstedt compares it to the great religious monuments of the past: "It takes its place with a handful of other preeminent buildings such as the Parthenon, Hagia Sophia, and Abbot Suger's St. Denis." Prince Albert, Queen Victoria's consort, would have agreed. According to him, the Crystal Palace is a

Joseph Paxton, Crystal Palace (1851). Illustration from Scientific American, *March 19, 1851.*

cathedral celebrating the goodness of the Creator: "Say not the discoveries we make are our own—The germs of every art are implanted within us, And God our instructor, out of that which is concealed, Develops the faculties of invention."

The essential fact about the Crystal Palace is that it was a response to an emergency. Its building commission had failed to approve any of the nearly 250 designs submitted to it and had failed to come up with an alternative. At this point, Joseph Paxton, gardener for the Duke of Devonshire and a self-taught builder of greenhouses, offered to design a structure that would satisfy the requirements for exhibition space and could be erected within the time available.

The commissioners would have preferred a design that alluded to history. One of the designs they favored initially was dominated by a cast-iron dome two hundred feet in diameter like the dome that would later be erected over the American Capitol. Paxton offered abstraction. He had to, because there was no time for symbols. Only the most advanced technology could create the building within the time available, and the structures produced by Paxton's technology were functional and abstract.

Erection of the building took a mere thirty-nine weeks. When finished it covered nineteen acres and used nine hundred thousand square feet of sheet glass. It was held up by some three thousand columns, and not one of them could be confused with Ionic or Doric columns. They were essentially tall, slender iron cylinders—pipes, really. Notably missing from Paxton's scheme was a monumental equivalent to the abandoned cast-iron dome or any other feature that might be interpreted as a concession to history—or, at least, to any history other than the history of greenhouses.

John Ruskin spoke for proper Victorians when he deplored the Crystal Palace. It was, he complained, a utilitarian structure lacking in grace and historical significance. A reasonable place to display things but "Neither a palace nor of crystal." Many contemporaries, however, were delighted by it. Lothar Bucher, an architectural critic, announced immediately after it opened: "A Midsummer Night's Dream seen clearly from midday." Jerome Buckley offers a modern appraisal in *The Victorian Temper* (1951): "The Crystal Palace, breaking all orthodox precedents, raised its airy shell, supported by a vertebrate structure of light blue iron girders, as a thin transparent cover . . . shaped . . . to suggest all the unlimited outer world by the space within."

The Crystal Palace created what might be called an aesthetic shock wave. It provided a rallying point for those who argued that architecture was too much concerned with ornament and not enough with technology. It also influenced the design of exhibition buildings in Dublin, New York, and Munich, and encouraged the use of glass in shopping malls like Milan's Galleria Vittorio Emanuele. One Victorian who visited the Crystal Palace, William Whiteley, was inspired by it to incorporate large plate-glass windows in his shops, and thus created a new urban institution that gradually replaced the emporia of an earlier era—the department store. There is evidence that the Crystal Palace encouraged the use of glass curtain walls by the architects of the Bauhaus. However, it had only a modest influence on buildings other than those designed for display and selling.

Evidently it was *too* functional—so obviously tied to its use as an exhibition building that its larger implications were invisible, even as the Great Wall of China must have been invisible to Shih Huang-ti's concubines as long as the barbarians were massed on the other side.

The Eiffel Tower presents a situation exactly opposite to that of the Crystal Palace. It was built in 1889 to commemorate the centennial of the French Revolution. Like the Crystal Palace, it was the product of a competition. Gustave Eiffel, its architect, was a bridge builder with a remarkable sense of the aesthetic inherent in technology. His Douro River Bridge in Portugal (1877) and Garabit Viaduct in France (1884) had already shown extraordinary flair for design. His later life was shadowed briefly by the debacle of the French Panama Canal, but he survived handily. He lived until 1923, and in the early years of the twentieth century he built a wind tunnel and experimented with airfoils.

There is nothing in the sweeping curves and intricately woven trusses of the Eiffel Tower to suggest the French Revolution it presumably commemorates. It is a work of pure engineering, a magnificent representation that represents only itself. As Roland Barthes remarks in a famous essay, "La Tour Eiffel," its sole function is to join its base to its pinnacle. By rejecting historical symbolism and the unique historical and cultural traditions of the city it dominates, it achieves the universality, the independence of space and time, of the technology that made it possible. It is a bridge rotated from horizontal to vertical and, at the same time, a fully realized abstract sculpture. In this sense it is a new word, an explosion of sound in an oppressive silence.

Above all it is there. Most bridges can be ignored for the same reason that the Great Wall could be ignored. They are useful devices, ways to get over a gap, but have no pretense to identities of their own. Their only message is "Cross me." But the Eiffel Tower cannot be ignored, simply because it has no use. You cannot cross it. The only thing you can do other than ride up its elevator and look *from* it is look *at* it. (It has a restaurant; presumably, you go to that restaurant not for the cuisine but so you can look from it while eating.) Ever since it was built it has been shouting at the top of its great iron lungs: "Here I am! Look at me!"

"Here!" for the Eiffel Tower was the center of the civilized world, Paris, the City of Light, the arbiter, at the end of the nineteenth century, of everything considered tasteful, brilliant, culturally vital—above all, aesthetic. When the Eiffel Tower raised its spire above Paris it shouted to *tout le monde*—the entire world, or all of the world that mattered—that the aesthetic of the abstract had arrived. It thus forced viewers to recognize something that pervaded and shaped their lives but had remained nameless. To recognize the "something" was not to discover the new but to see the face of the modern era—one's own face—in the mirror.

Gustave Eiffel, Eiffel Tower (1889).

When the leaders of the French establishment heard the cry of the Eiffel Tower in 1889 they were outraged. A protest signed by Alexandre Dumas and Guy de Maupassant, among others, expresses the shame they felt at the thought that they would have to listen to its message for the rest of their lives:

> We come to protest with all our strength, with all our indignation, in the name of disregarded French taste, in the name of art and French history presently in danger, against the erection, and in the very heart of our capital, of the useless and monstrous Eiffel Tower. . . . Will the city of Paris continue to listen to the baroque, merchantile fancies of a builder of machines, and irreparably lose its honor and beauty? For the Eiffel Tower, which even the commercial America would reject, means, without any doubt, a Paris dishonored.

The reference to "commercial America" allows an enlargement of the lesson of the Eiffel Tower. Thirteen years before it was built, the French decided to make a gift to America to commemorate the 1876 centennial of the Revolutionary War. Perhaps the sponsors of the project had talked about American taste with Dumas and de Maupassant. They do not seem to have considered a tower. Instead they decided to give America a work of art. They appointed A. A. Bartholdi, a sculptor, to make one: a female figure draped in ponderously diaphanous metal robes. She is pure symbolism. She wears a crown of light and holds a torch of freedom aloft to guide oppressed masses into New York harbor. She is a Roman goddess. Liberty. A sister of the goddess *Freedom* who stands atop the dome of the Capitol. *Quel beau sentiment!*

Because Liberty was so large, she required considerable engineering. A skeletal frame was needed to support her. Given her size and the weight of her garments, that was no small task. The more difficult task was to find a way of attaching the plates that make up Liberty's body and robes in such a way as to allow them to expand and contract without falling off.

The details were worked out by none other than Gustave Eiffel. Eiffel's engineering is ingenious but it is hidden like genitalia under Liberty's robes. It is an embarrassment, at best, a necessary means toward an end. The end has no more to do with technology than the structure of the Brooklyn Bridge has to do with a plan to drive to New York to see a movie. Liberty means something that has been defined by history. She has a name.

The Eiffel Tower is, conversely, all engineering and no robes. Being abstract—devoid of historical allusions to the French Revolution—it can be read as a symbol of almost anything: a prophecy of flight, an expression of man's aspiration for the infinite, a phallic symbol, an enlarged toy. It is none of these. It forces the viewer to look beyond historical myths to the revolution that surrounds him. It does not argue with history, it ignores history. It is as though the president of the French Republic appeared at a state dinner wearing Levi's.

Dumas and de Maupassant raged, but the French souvenir business prospered. After the Eiffel Tower, who could be satisfied to leave Paris with a metal paperweight shaped like the Arc de Triomphe?

What tourists knew by instinct avant-garde artists would soon discover: the Eiffel Tower is the first great icon of the technological aesthetic. Two others are John Roebling's Brooklyn Bridge and Vladimir Tatlin's Monument to the Third International. Both were celebrated by artists almost as fervently as the Eiffel Tower, although Roebling's bridge mixed historical allusions—Gothic arches, for example—with its technology, and Tatlin's monument was never built.

The first artist of the Eiffel Tower was Robert Delaunay, who painted it some thirty times. "The Tower," he said, "addresses the universe." And so it did. After 1909 it was used for radio transmission. A friend, Vincente Huidobro, dedicated a poem entitled "Eiffel Tower" to Delaunay in 1917. In translation:

> Eiffel Tower
> Sky guitar
>
> Your wireless telegraph
> Pulls words toward you
> Like a rose arbor attracting trees. . . .

No wonder the leaders of *fin de siècle* culture were apoplectic. But rejecting history has its uses. Thomas Jefferson wrote to his friend Cartwright, "The Creator has made the earth for the living, not the dead." The Eiffel Tower announces that the twentieth century will be for the living, not the dead.

10

BAUHAUS

In 1919, a cooperative school of architecture and design led by the German architect Walter Gropius assumed control of the Grand-Ducal Saxon Academy of Art in Weimar and promptly gave it a new name—Bauhaus. The event marks a watershed in the development of modern consciousness fully as significant as the publication of Einstein's paper on general relativity in 1905.

Before Gropius, the Grand-Ducal Academy had insisted that the arts are interrelated and that too much specialization impoverishes both. The Bauhaus extended and strengthened this idea by insisting that all of the arts must be understood in relation to architecture. The architect produces all-encompassing concepts with which the other, more specialized art forms must be coordinated. A painting, for example, makes complete sense only when it is understood in relation to the space in which it is to be hung. "The separate arts," wrote Gropius in 1919, "must be freed of their isolation and must be brought back into intimate contact, under the wings of a great architecture."

The Grand-Ducal Academy had also been committed to revitalizating the arts and crafts movement that flourished throughout Europe before the First World War. The crafts movement was a movement *against* the impersonal forces of mass production. It romanticized the image of the medieval craftsman lovingly and anonymously producing wood carvings, tapestries, and ornate ironwork. The first (1919) prospectus of the Bauhaus implies that it has the same ideal. Totally absent are suggestions of a plan to find points of contact between quality design and the mass-production techniques of twentieth-century industry.

A different credo had, however, been formulated by Gropius as early as 1910. At that time he was already arguing that the age of the individual craftsman had vanished forever: "Only through mass production can really good products be provided. . . . The trend of our age to eliminate the craftsman promises far greater industrial rationalization." His sensitivity to technological aesthetic was clearly encouraged by the industrial architecture he produced before the First World War. His Faguswerk building (a factory for making shoe lasts) in Alfeld-am-Leine was built in 1911. It is severely rectilinear with cantilevered glass curtain walls very much in the later Bauhaus style. Today it is a German national landmark.

In 1916 Gropius identified the new aesthetic with industrial products: "Accurately cast forms, bare of ornamentation, having clear contrasts and consistency of form and color, become commensurate with the energy and economy of modern life."

Johannes Itten, one of the original master teachers at the Bauhaus, was a strong advocate of the crafts concept. In 1922 he challenged his colleagues either to commit themselves to the crafts concept or to state openly that they intended to cooperate with industry. Gropius responded by formulating a program for the Bauhaus fully committed to the needs of technological culture and to the idea of "a well-built automobile, an airplane, and a modern machine as individual works of art beautifully formed by creative hands."

Before long Gropius was explicitly condemning the crafts ideal and arguing that the preoccupation of the crafts movement with history was empty posturing: " . . . with very few exceptions, [it is] a lie. In all of these products one recognizes the false and spastic effort to 'make Art.' " By contrast, "The engineer . . . unhampered by aesthetics and historical inhibitions, has arrived at clear and organic forms. He seems to be slowly taking over the role of the architect, who evolved from the crafts."

An exhibition in the summer of 1923 publicized the new technological-abstract focus of the Bauhaus. The central object was a model house called Am Horn, in which an attempt is made to integrate all of the elements of the Bauhaus program as architecture, design, and decoration. Am Horn looks small, fragile, and a little drab in the black-and-white photographs that survive. One is inclined to ask, Why all the excitement? How could something so humble lead to such an enormous revolution?

The house is a one-story box surmounted by a second, smaller box. The most obvious feature of its aesthetic is its emphasis on squares, rectangles, and right angles. The roof is flat. The second, smaller box with clerestory windows rises in the middle over the central living room. The exterior walls are white stucco applied over prefabricated "plates"—an attempt to reduce the hand labor involved in brick construction. The interior walls are plain, and doors and windows are flush. The kitchen is notable for its continuous work counters and cabinets. Special care is evident, too, in the design of the bathroom, which has white glass-ceramic walls and a rubber tile floor.

Am Horn House (Bauhaus, 1923). Note emphasis on geometric regularity, especially on right angles and on strong horizontal and vertical lines.

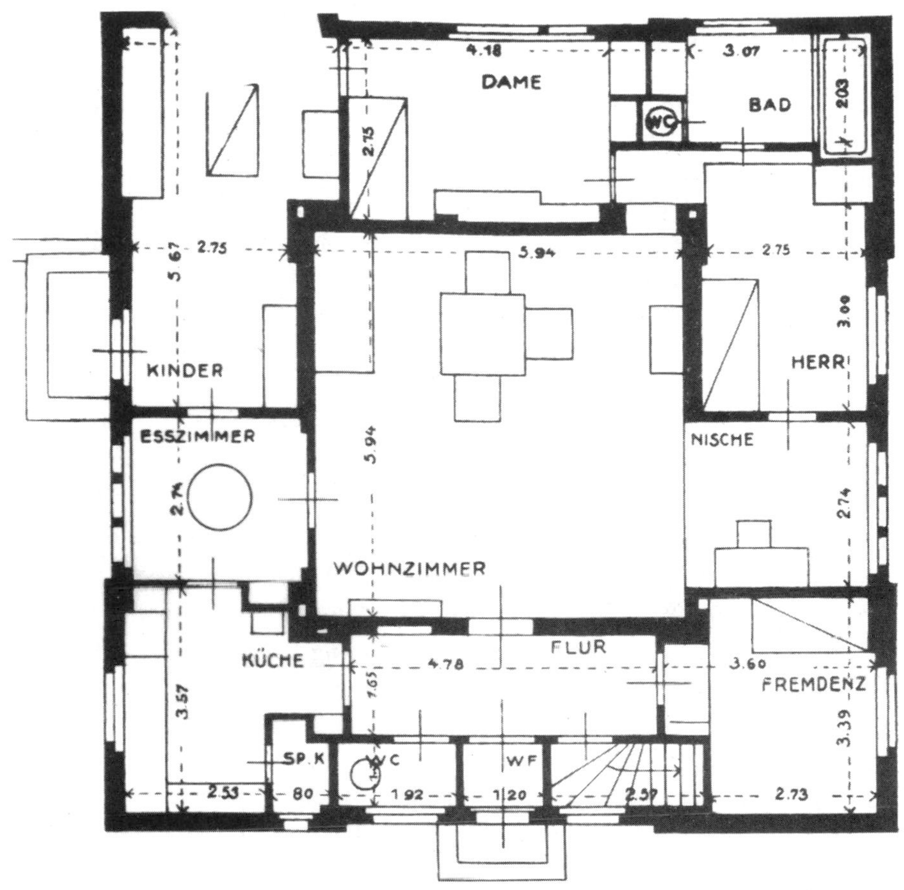

Floor plan of Am Horn House (Bauhaus, 1923). Note the sheltering of the central living room.

Because the house is understood as a refuge, the windows are placed to introduce light rather than to provide views of the outside world. The light plays against a generally muted color scheme. A child's room has red, blue, and yellow wainscoting and features a blackboard for drawing and modular furniture that can also be used as large building blocks. Throughout, closets are built in, and lighting is recessed and filtered through tinted tubes. The built-in feeling, while not unique to Bauhaus design, enhances the uncluttered and functional feel of the interiors.

A manifesto was included in the pamphlet publicizing the exhibition. It noted the relationship of technology, cost-effective construction techniques, and abstraction:

> Mathematics, structure, and mechanization are the elements, and power and money are the dictators of these modern phenomena of steel, concrete, glass, and electricity. Velocity of rigid matter, dematerialization of matter, organization of inorganic matter, all these produce the miracle of abstraction. Based on the laws of nature, these are the achievements of mind in the conquest of nature. . . . The speed and supertension of commercialism make expediency and utility the measure of all effectiveness, and calculation seizes the transcendent world: art . . . long bereft of its name, lives a life after death, in the monument of the cube and in the colored square.

The interest of the Bauhaus in rectilinear shapes was shared by Theo van Doesburg, Piet Mondrian, and Georges Vantongerloo, leaders of the Dutch movement known as De Stijl. Mondrian was passionately devoted to formal purity and in particular to rectilinear form, which he believed was a means of revealing the spiritual force that is at the center of reality. He explained in *New Design* (*Die Neue Gestaltung*), published by the Bauhaus in 1923, that abstraction in painting is continuous with the geometries of Bauhaus architecture:

> The new aesthetics of architecture is the same as that of painting. And building, which is in the process of clarifying itself, is already putting into effect the same findings that painting realized in the "new design" after a process of clarification heralded by Futurism and Cubism. Because of the unity of the "new esthetics," building and painting can constitute *one* art and mutually absorb one another.

Visitors to Am Horn did not need Mondrian to discover that right angles were, indeed, in. A reviewer of the 1923 exhibition remarked, "Three days in Weimar and one can never look at a square again for the rest of one's life."

Although it is modest in size, Am Horn is a decisive statement. It is a summation of design philosophy that could be presented only through a house because it conceives of design as a total environment. Windows, wall textures, cabinets, millwork, lighting fixtures, paint colors, floor surfaces, draperies, and utensils are to be harmonized in terms of the overarching architectural program. Am Horn is not a model for a possible suburban development. It is the manifesto of a program to redesign the human environment.

In 1925 the Bauhaus moved from Weimar to Dessau. The move required the creation of a "workshop building." At the ceremonies opening this building Gropius took justifiable satisfaction in what had been accomplished. The ideas of the Bauhaus had achieved international recognition. Now they were codified in a large-scale building rather than a small private house. Like Am Horn, the new building brought together ideas later recognized as characteristic of the Bauhaus. It had three wings. Each had unique design features reflecting a special function. Throughout, however, simple shapes were emphasized. The façades were strongly rectilinear, with emphasis on the play of vertical against horizontal lines. The roof was flat, its extensive surfaces protected by an experimental membrane. As luck would have it, the membrane leaked. The leakage was repeatedly cited by opponents of the Bauhaus as evidence that Gropius and his associates were incompetent.

To allow maximum light, the studio section used curtain walls of glass with panes segmented into rectangular units by steel mullions cantilevered outward from supporting piers. A reviewer from Berlin noted the enchantment of the effect: "The glass which separates [inside and out], and then half cancels this separation, envelops the giant cube of the building with a transparent skin through which one can see the pulsating activity on the inside of the organism. This is twice as fascinating at night when the cube is lit up and the surroundings are bathed in the blinding light from its interior."

The style popularized by the Bauhaus was reaffirmed in 1930 when Ludwig Mies van der Rohe, the last of its directors, was appointed. In what might be considered a summing up of the lessons learned in the 1920s, he remarked, "The new age is a fact; it exists entirely independently of whether we say 'yes' or 'no' to it." The Nazis said "no" to the idea of a new age and proceeded to attack the Bauhaus. By 1933 they had driven it to Berlin and killed it. Mies van der Rohe said "yes," and the price he paid was being forced into exile. With the assistance of Philip Johnson, then an architectural historian, he would resume his career in Chicago. Gropius went to Harvard. Josef and Anni Albers went to Black Mountain College in North Carolina. Laszlo Moholy-Nagy eventually came to Chicago to found the Chicago School of Design, which survives today as a division of the Illinois Institute of Technology.

11

INTERNATIONAL STYLE

In 1932 the Museum of Modern Art in New York held an exhibition of modern architecture which was largely a tribute to the Bauhaus. In the catalogue for the exhibition, Henry-Russell Hitchcock and Philip Johnson identified the style the exhibition revealed. They called it the International Style.

The phrase was a stroke of genius. It expressed the universalizing thrust—the push away from local attachments—of the new architecture. Like so many avant-garde artists of the twentieth century, the Bauhaus architects believed they were helping to usher in an era of world culture. Although the features of this culture were appearing first in technologically advanced societies, principally Europe and North America, the new culture would be transnational and transregional. An architecture suited to it would share these qualities and express them. It would be international because it would draw on the aesthetic of technology, not the grab bag of styles produced over the centuries by the different historical circumstances of different countries and regions.

It would also serve the ends of social justice. Perhaps a few of the pioneers of the International Style were interested only in money, but most shared the ideals of the founding members of the Bauhaus, who believed they could elevate modern culture, especially the culture of the working class. The Nazis called them anarchists and Bolshevists, but their idea was a peaceful transformation of society through designs that would touch every class because they could be reproduced cheaply and spread widely by mass production. In the process, by creating a world culture they would contribute to the reduction of age-old hostilities and tensions.

It is important to recognize these social aims because the argument is often made that the International Style became a rubber-stamp method of design—what the Germans call *Stempelarchitektur*. In one sense the charge is just. The strategies of the Bauhaus were intended from the beginning to be simple, practical, and cost-effective, which means, in an industrial context, capable of being reproduced cheaply and easily. One man's *Stempelarchitektur* is another man's mass-production housing. If the strategies of the new architecture were not better than the old from the point of view of engineers and economists as well as from the point of view of aesthetics, how could it succeed in its mission of transforming society?

The fact that the International Style was enormously successful, the fact that it became the standard design for an enormous range of buildings from the most elegant to the most slipshod, is an index of the practicality of its original concepts.

Between 1945 and 1970 the International Style swept all competition aside. Hotels, motels, office buildings, factories, museums, private houses, department stores, hospitals, low-cost housing developments—in short, every imaginable sort of building—testify to its impact. The International Style is not only international in the sense of being abstract rather than historical, but also in the sense of being worldwide. Low-cost housing and public-housing projects from Singapore to Hong Kong to Hamburg to Chicago demonstrate that it can be cheap and modular as well as expensive and heroic. It can produce clean, bright, efficient one-bedroom apartments that recall Am Horn as well as airy Pythagorean fantasies.

The best known American examples of the International Style are strongly geometric and heroic in size. These buildings are examples of high-cost, high-finish International Style, but even expensive buildings can have a social purpose because wherever they rise in the urban environment they teach their aesthetic. This is an important point. They may simply seem functional to the casual observer but they have an aesthetic mission. Three New York buildings that show how powerful the aesthetic element can be are Gordon Bunshaft's Lever House, Mies van der Rohe's Seagram Building (with assistance by Philip Johnson), and the U.N. Secretariat building designed under the direction of Wallace K. Harrison.

Mies's Seagram Building (1958) is an especially striking example. It is elegantly functional in appearance, but the functionalism is to a certain extent an illusion. The building seems to reveal its structure by exposing its vertical I beams. Is this not a clear-cut case of letting structure speak for itself? Actually, the beams are a kind of visual comment on structure rather than structure proper. They have no function except decoration. The real girders are inside the building and invisible, encased in the fireproof cement required by New York's fire code.

Silver and bronze and black reflective glass covers many buildings in the late International Style nicknamed "Los Angeles Silver." In this form the Pythagorean geometries of the Bauhaus experience a rebirth and present themselves with a breathtaking purity that Mies van der Rohe himself could not fail to admire.

The use of reflective glass is functional, especially in the American Southwest and South, because it reduces heat radiation into the buildings. However, the primary objective is beauty, not efficiency. In good buildings the effect simplifies the structures, getting rid of the busy reticulations of earlier International Style buildings. The squares and rectangles are replaced by gigantic mirrors reflecting surrealistic variations on the surrounding activity (see illustration 5 in color section).

Human beings float across these mirrors like bubbles in an oversized Wurlitzer jukebox. The buildings themselves seem only half real. Sometimes they are brilliantly there—great rectangles and cylinders of re-

Mies van der Rohe, with Philip Johnson, Seagram Building, New York City (1957).

flected light. At other times they are almost invisible because when they are looked at they blend into the scenery.

When they are not disappearing, these buildings assert by their geometries that they are rigid and that their basic structural principle is compression, the piling of one column on another. The movement that ripples across their surfaces intensifies the impression that they, themselves, are static.

To many, the defining buildings of the International Style are breathtakingly lovely. Others consider them abominations. In *From Bauhaus to Our House* (1982), Tom Wolfe attacks the International Style for producing "glass boxes" and "great hulking structures," which even those who commission them detest. In total indifference to the politics of its founders, critics of the International Style, especially advocates of postmodernism and urban restoration, have called it inhuman and "puritanical" and "fascist." Its emphasis on undecorated abstract form has been compared to "an Arizona of the mind."

The ideal of making good design available to the people at large at modest cost is, you might think, the opposite of fascism. Much more important, the better products of the International Style both objectify the imperatives of technology and proclaim the technological aesthetic. Is not an admiration of grand and simple geometric forms closer to the quasi-religious idealism of D'Arcy Thompson than to "an Arizona of the mind" or a soulless formalism? Such forms are simultaneously abstract, un-natural, and universal, and thus proclaim confidently the ideal of an international culture freed at last from the distractions of local history and parochial tradition.

The success of the International Style is not devoid of irony. It became a world architectural style precisely because it worked. It *could* be duplicated. Yet the proliferation of buildings using the style is the cause of most of the attacks on it. If the International Style were less commonplace, if it were less capable of being made into *Stempelarchitektur*—if, in other words, it were elitist in the manner of the palace architecture of preindustrial cultures—it would doubtless have caused less of a reaction. A 1978 critique by Charles Jencks of Philip Johnson's AT&T building in New York praises the building for its rejection of International Style. The reason is unabashedly elitist. Johnson's building leads Jencks to hope that the skyscraper will "lose its bland economic coding and

llace K. Harrison
l others, U.N.
retariat building,
w York City
47). A statement
the universalist
itical ideals as well
the aesthetic of the
ernational Style.

return to its former position as a major fantasy form of capital (whether capitalist or socialist)."

The U.N. Secretariat building in New York combines its aesthetic with the statement of a political ideal. For the statement to be understood, the Secretariat must be contrasted to the Capitol Building in Washington. The Capitol is the metaphysical center of American space. Its classical decor celebrates an American political tradition that extends backward in time to the Age of Pericles in Greece and of Cicero in Rome.

The Secretariat is the expression of the transnational and transhistorical ideal that led to the founding of the United Nations. A rectangular slab 544 feet high but only seventy-two feet from back to front, it is pure geometry, as free of history as a Mondrian abstraction, and it announces with an assurance that has all too often seemed absurdly misplaced in the years since its completion that it is the capitol of planet Earth. Perhaps in the Gorbachev era of international relations the assurance will become less problematic. Whatever the case, the space the Secretariat building has in mind is global and the time is now, and the ideal it expresses is a fulfillment of the internationalism of the leaders of the Bauhaus movement.

12

LET'S PLAY ARCHITECTURE

The John Hancock building in Chicago, designed by Skidmore, Owings, and Merrill, was completed in 1969. It recalls the fondness of the International Style for simple geometries. In the contest between structure and form, however, structure has asserted itself. It is not a serene geometric shape. It tapers as it rises from its powerful base.

Technically it is a truncated pyramid, but it does not permit the eye the satisfaction of the pyramid form. The taper is too slight and the top is too emphatically a flat roof. The taper therefore insists on being understood structurally as a strategy to reduce sway. The structural interpretation is reinforced by the aggressive exposure of diagonal bracing that zig-zags its way up the sides of the building.

The John Hancock building is one of the great buildings of the twentieth century, but it is also a departure from the formalism of the International Style. Although the Bauhaus was influenced by the no-nonsense functionalism of nineteenth-century industrial structures like silos and sheet-iron warehouses and railway bridges, it moved beyond these to elegant geometric simplifications that soon developed their own aesthetic. The overt display of structure that is so striking in the John Hancock building is a reversion to the functional aesthetic in its elemental form—a form entirely congenial to the native traditions of Chicago. As Hans Wingler recalls in his history of the Bauhaus, "The great architects who assembled after the Chicago fire of 1871 . . . were able to turn the achievements of the engineers to their own purposes. . . . It was in this presentation that functional architectural thought . . . found its first consummate expression."

The John Hancock building is an early expression of dissatisfaction with geometric formalism that is expressed far more directly by post-modernism, where it is associated with a much advertised return to history.

Here, the influence of Philip Johnson, architectural historian-turned-architect, is central. Johnson began his career with the catalogue of the 1932 exhibition at the Museum of Modern Art celebrating the work of the Bauhaus. In 1949 he built a glass cube for himself in New Canaan, Connecticut, that became a notorious symbol of the invasion of America by the International Style. Later, he would collaborate with Mies van der Rohe on the Seagram Building.

Yet by 1975 Philip Johnson had become one of the most visible American critics of the International Style. An early phase of his apostasy is illustrated by the Pennzoil building in Houston (1976), which rejects rectangles in favor of a slashing diagonal geometry for the upper stories. Johnson's term for the style is "shaped modern." A more descriptive name would be "beveled style." The style has produced endless variations on the theme of rectangular solids with surprisingly angled edges and corners and indented surfaces, and it has also encouraged free use of pyramids, prisms, rhomboids, and the like. The beveled style is often striking, but it remains emphatically geometric and in this sense a style free of traditional historical symbolism.

The large cylindrical buildings that became popular in the 1960s are

Skidmore, Owings, and Merrill, John Hancock building, Chicago, Illinois (1969). Note the building's taper and the exposure of the powerful diagonal bracing structure.

another significant variation on geometric formalism. Cylindrical high-rise buildings were sketched by Mies van der Rohe in the 1920s, but the cylinder did not become commonplace in architecture until it began to be used in America for offices and hotels. Two defining examples of cylindrical structure are the Peachtree Plaza Hotel in Atlanta and thc Renaissance Center in Detroit. The popularity of the cylinder is due in large part, one feels, to the fact that it is a beautiful shape both in itself and in association with other simple geometric forms. When large, cylindrical buildings are sheathed in reflective glass, they take on the illusionary qualities already noted in mirror-sheathed rectilinear buildings.

Buckminster Fuller's geodesic domes illustrate another direction taken by geometric architecture since the 1950s. Their geometry is intricate, like the geometry of a Tinkertoy, not simple, but it is powerfully evident to anyone who sees them.

Postmodernism does not offer itself as a variation on the International Style but as an alternative. It promises exactly what the International Style has rejected: decoration for the sake of decoration and the use of historical symbolism through what theorists of postmodern style call "quotation"—that is, details and design motifs taken from buildings of the past. A building that includes part of a colonnade might, for example, be said to "quote" the colonnade surrounding St. Peter's Square in Rome designed by Giovanni Bernini; a series of rowhouses in a semicircle might be said to "quote" the Royal Crescent designed by the two John Woods of Bath, England.

When quotation is serious, it is understood historically. The columns on the Capitol Building, for example, can be called quotations of Roman architecture, and they are undoubtedly to be taken seriously. The point is that in postmodern architecture, quotation is usually more lighthearted. It tends either to the fantastic—to artificial, dreamlike structures—or to the ironic—that is, structures that combine divergent or clashing traditions. It is thus not an expression of historical piety so much as a rejection of the geometric tradition of Cubism in favor of the eclectic tradition of collage. In its indifference to deeper historical meanings and its fondness for irony it is consistent with the twentieth-century habit of collocating materials from diverse, often conflicting traditions.

The Piazza d'Italia in New Orleans by Charles Moore, one of the authentic prophets of post-modernism, exemplifies architectural fantasy. It is an evocation of Italian Renaissance architecture created by quotation of every imaginable cliché of *cinquecento* design—complex trick perspective, arches supported by columns, apse, Piazza di Spagna–like

(Left) Mies van der Rohe, Model for a Glass Skyscraper *(1921). An early experiment with the cylinder in contrast to the rectilinear geometry more typical of the Bauhaus and the International Style.*

(Below) John Portman, Peachtree Plaza Hotel, Atlanta, Georgia (1976). An influential example of the use of the cylinder in American commercial architecture.

stairway, lagoonlike pond, curved colonnade–like wall, monumental inscriptions in raised capital letters, and more. The intended result is not history but play. Moore's Piazza is a cousin of the fake national environments—Chinese, Norwegian, Mexican, and so forth—at Disney World's Epcot Center outside of Orlando, Florida. Epcot Center was explicitly designed as a playground. You do not have to travel to New Orleans or Orlando, however, to see fantasy complexes. They are standard features of the enclosed shopping malls that dot the suburban landscape of every middle-sized American city.

Robert Venturi's *Complexity and Contradiction in Architecture* (1966) is one of the key manifestos of postmodernism. Venturi announces, "Architects can no longer afford to be intimidated by the puritanically moral language of orthodox modern architecture. I like elements which are hybrid rather than 'pure,' compromising rather than 'clean,' distorted rather than 'straightforward,' ambiguous . . . rather than direct and clear. I am for messy variety over obvious unity." It is a stirring message that clearly places more emphasis on ambivalence and collage than on history. It points to fantasies like Moore's Piazza rather than to careful historical evocations like the Capitol Building. Venturi enlarged his message in a book published in 1972: *Learning from Las Vegas*. In *Learning* architecture is urged to incorporate colloquial forms like those found on the Las Vegas strip. Quotation from these forms helps to make design "hybrid" and "ambiguous."

The large-scale equivalent of the hybrid and ambiguous effects of postmodern buildings is architectural pluralism. This involves hostility to city planning. Charles Jencks explains the position in a now-classic essay entitled "The Rise of Postmodern Architecture," later expanded to *The Language of Post-Modern Architecture* (1977). Two of Jencks's key terms are "adhocism" and "radical traditionalism."

As he explains, the ad hoc city is intended to avoid the horrors of Le Corbusier's idea of the totally planned city. Corbusier is often depicted by postmodernists as the father of such well-intentioned but disastrous efforts at planned low-cost housing as I. Yamasaki's Pruitt-Igoe project in St. Louis, which was such a disaster of planning that it was blown up by the city in a spectacular and much-photographed demolition on July 12, 1972. But expensive and elaborately designed projects are equally suspect, a frequently cited example being the city of Brasília, the capital of Brazil, which is regularly depicted as monumental, inhuman, out of phase with the real lives of those who live in it, and, increasingly, a mixture of sterile geometries and squalid slums. The implication is that the sad state of Brasília is the direct result of its architecture rather than the depressed state of Brazil's economy and the deficiencies of its city services.

Adhocism is everything that planning is not. According to Jencks it "feeds off problems like a glutton. It celebrates the collision of subcultures, invents the botched joint, and cheerfully slams different options together." It is also enthusiastically on the side of urban preservation. Apparently, everything that exists should be preserved. Jencks argues, "There is no real economic or structural reason for destroying any building—just the commitment to harmony, the clean slate, and the bulldozer ethic."

Preservation should not, however, be confused with the preservation of history. In Jencks's view it creates illusion: "*trompe l'oeil* urban scenery." This comment is interesting because it recognizes that what is wanted is not history in the sense of a return to tradition and a stable sense of identity, but fantasy. The ideal of the ad hoc city clearly echoes the fondness for clashing styles and ironic "quotations" of postmodern buildings as well as the creation of explicitly fantastic architectural complexes.

In *Collage City* (1975) Colin Rowe calls for a city that is a rich mixture of styles. The collage idea might be considered historical because it implies the preservation of many bits and pieces of history. In the urban setting mixed styles always teach history lessons. Rowe is not, however, very interested in history lessons. He likes collage city because it is fantastic and playful—in other words, because it offers an escape from history. His ideal seems to be something like an urban equivalent of Charles Moore's Piazza d'Italia. He calls the architect a *bricoleur*, a snapper up of unconsidered trifles. The *bricoleur* deals with "a plurality of closed finite systems—*the collection of oddments left over from human endeavour.*"

For advocates of postmodernism, the completion in 1978 of Philip Johnson's AT&T building on Madison Avenue and Fifty-sixth Street in New York was a moment of truth. Turning away from the beveled modernism of Pennzoil Place, Johnson created a thirty-seven-story building with an enormous arched entranceway reminiscent, according to some reviewers, of the seventeenth-century Roman entrances of Francesco Borromini. The building face is granite (though with powerful vertical lines) rather than glass. The granite reinforces the pretensions of the entrance by suggesting monumentality.

As a matter of fact, monumentality is a dominant motive of Johnson's building. Borromini and other Renaissance architects were careful to observe the ancient sense of human proportion at the same time that they quoted classical motifs. In the AT&T building, monumentality is its own excuse for being. The result is that its allusions to earlier architecture are at war with themselves. They do not express the humanity of architecture; instead they announce a kind of disappearance of humanity from their frame of reference.

The triumphant feature of the AT&T building is its roof. The roof is not flat, as in the typical high-rise buildings of the International Style. Its front is a pediment—a genuine historical quotation. But of what? A

Philip Johnson, AT&T building, New York City (1978). In many respects, the manifesto of the postmodern movement in American architecture.

Roman temple? Surely not, because there is a large hole in the middle of it. The hole is intended, Johnson assured the critics, to puff smoke from the building's heating system in cold weather. Smoke or not, the quotation was recognized immediately by all critics. The roof of the AT&T building is a Chippendale quotation. Does it quote a clock or a chest of drawers? Or is it an ironic quotation—say, of a 1925 Chippendale table radio? Clearly, it is not a sober tribute to Roman history like the goddess *Freedom* atop the Capitol dome. The quotation is an elephantine joke. It may not double you over, but you cannot look at it without smiling.

Postmodernism is mildly odd but its ironies can be appealing. They depend, like ironic literary quotations, on control by the artist.

In other forms of modern architecture the artist surrenders some of that control. The result is not a deeper humanity but an increasing sense of architecture as mosaic.

In mosaic architecture, history is not ignored as it is in the International Style. Nor is it quoted by architects in relation to larger design concepts which they control, as in postmodernism. Instead, it combines bits and pieces of older buildings with bits and pieces of modern buildings. The combining is explained in the most pious terms—it is a way to save architectural treasures that would otherwise be destroyed. However, quite apart from the fact that the treasures are often banalities, they are often saved only to be demeaned. The brutal contrast of the Dirksen Senate Office Building with the Hart Building anticipates the trend, although in that case the effect is created by juxtaposition rather than combination.

A common form of mosaic architecture is façadism, in which the fronts of nineteenth-century buildings are propped up while entirely new buildings are created behind them and (often) beside and above them. A bizarre example—and perhaps the most vulgar structure created in America between 1965 and 1985—is the Red Lion Row development in Washington, D.C., five blocks from the White House. This structure—if structure it can be called—appears to be a block of three-story rowhouses restored with such germicidal precision that it looks like a stage set. That is exactly what they are. Behind them rises the real building, an immense, smooth, ominous steel-and-glass structure that stares down on them with equal measures of contempt and surprise, like the owner of a penthouse examining cockroaches on the kitchen floor.

Vulgarity has its charms, and mosaic architecture has its own aesthetic.

J. Carl Warnecke, Red Lion Row redevelopment, Washington, D.C. (1982). Photograph by Stavros Moschopoulos/Image Ray.

The aesthetic is not, however, derived from a reverent evocation of history. It can be an aesthetic of irony, but more often it is closer to an aesthetic of randomness, and sometimes it is the aesthetic of discontinuity or of contradiction. Still more bizarre are the designs of James Wines of SITES for Best and Company stores. Some of these are designed as ruins, some as monstrous shapes rising from beneath mantles of asphalt, and some as parodies of historical buildings. These structures have been highly effective in attracting the attention of shoppers, which is, after all, what stores are supposed to do. They are also a step forward beyond Red Lion Row in the undercutting of the historical element in architecture.

Urban historic preservation is reinforced by zoning laws creating historic districts and covenants to prevent stylistic collisions of the sort advocated by Colin Rowe. The apparent objective—like the objective of saving rowhouse façades—is to preserve history, but the more preservation succeeds in a given urban district, the less historical the result becomes. As the restoration moves forward, the district increasingly resembles a stage set. In a fully restored district, which is almost infallibly announced by epithets like "old," as in Old Santa Fe or Old Salem, and nouns like "village," as in Dearborn Village, history becomes a fantasy, a *trompe l'oeil* illusion, to quote Charles Jencks. The most successful of these districts have the same quality as intentionally fantastic creations like Disney World's Epcot and Charles Moore's Piazza d'Italia.

The logical limits of fantasy are museum villages and theme parks. In such environments everything is stage set and nothing is real. The visitor to Disneyland in Anaheim, California, enters along a street that looks like a meticulous reproduction of the main street of a typical nineteenth-century American town. However the effect is all illusion. The streets are unnaturally tidy, and the buildings are slightly under scale. The scaling is so subtle that the visitor may not be conscious of it, but it has a powerful psychological effect. The houses are oversized dollhouses—not quite real. The effect is at least as appealing to adults as to children. It is a dream of childhood freed of all conflict, pain, and ugliness.

Alternately, modern architecture can carry forward the functionalist tendency of Chicago's John Hancock building. This is the direction taken by Richard Rogers and Renzo Piano in the design of the Pompidou Center in Paris, also called the Beaubourg Museum.

The Pompidou, which opened in 1977, expresses its radical functionalism by exhibiting all of its mechanical services openly. Their bright colors and contrasts of shape, scale, and materials make their anti-aesthetic aesthetic explicit and continuous. Form follows function with relentless literalness. On the building's façade the most striking feature is an escalator that jogs its way up the outside wall with a pause at each floor. Visually prominent posts and tie-rods help support the building and maintain rigidity in ways vaguely reminiscent of the Crystal Palace. Structural features inside the building remain visible through the glass skin because there is no attempt to conceal them.

Aside from fire walls, the interior spaces are open and neutral. Supporting beams and other supporting structures are exposed. The theory behind this approach is similar to the theory underlying the construction of large office buildings. No one can predict what functions the building will be required to serve, and whatever functions it serves, these are sure to be replaced by others. Here, nonplanning is a functional design principle and history is precisely the impediment that the nonplanning is

Richard Rogers and Renzo Piano, Pompidou Center, Paris, France (1977)—beyond postmodernism.

intended to circumvent. The interior spaces face the future, not the past. So did the first director of the Pompidou Center, Pontus Hulten, who advertised his commitment to his building's architecture by riding around Paris on a motorcycle. When visiting the elegant marble East Building of Washington's National Gallery of Art he boasted, "Your museum seems a temple. Ours has the spirit of a factory instead."

One step beyond the Pompidou and you encounter neomodernism, which is also called deconstructive architecture. Neomodernism rejects functionalism in favor of delight and surprise. It is a playful architecture, a game. Like the International Style, neomodernism is an extension of radical experiments of the early twentieth century, but its inspiration is Russian Constructivism rather than Bauhaus formalism. Tall among its heroes is Vladimir Tatlin, designer of the grandiose and never-constructed Monument to the Third International, a twisted, spiraling tower that looks like a Braque staircase.

Vladimir Tatlin, model of proposed Monument to the Third International (1919). Tatlin was a leader of the Constructivist movement in Soviet art and architecture in the decade following the Russian revolution. The monument was never constructed.

We come, thus, to immediate and urgently active figures in modern architecture—architects like Zaha Hadid, Peter Eisenman, Frank Gehry, and Bernard Tschumi. In an article on neomodernism by Joseph Giovannini, Hadid explains that the movement is a carrying forward of the Constructivist search for stylistic languages that match the realities of modern life: "The most exciting thing about the Russians [the Constructivists] is that their experiment was never finished. . . . There was a certain dynamism in their work, and in how it affects plans. My notion of flight, of liberating the ground . . . developed from that. We can't carry on as cake decorators; we have to take on the task of investigating modernism, and carry this tradition into the 80s and 90s."

Two Viennese architects, Wolfgang Prix and Helmut Swiczinsky, appear to borrow concepts directly from fractal geometry. They intentionally introduce randomness into the design process and draw on "techniques used by geometricians who study the irregular patterns in apparently random natural phenomena, like clouds and coastline formations." The comment identifies the source of the theory. Who but Benoit Mandelbrot insists on introducing random variations in mathematical procedures and boasts of having created a "fractal geometry of nature" that can represent the shapes of clouds and coastlines?

Meanwhile, in the Parc de la Villette, a Parisian housing development, oddly shaped houses—the so-called follies of Bernard Tschumi, now dean of Columbia University's School of Architecture—blaze with primary colors, astonish with their unexpected angles and projections, and insult purists with their utterly useless decorations. They exhibit a kind of Heisenberg indeterminacy. At their most bizarre they are as surprising as the nature that modern science has revealed to us. The message they convey is that the modern experiment is not lagging. It has hardly begun.

13

THE BRIDGE AND *THE BRIDGE*

Georges Vantongerloo, the most brilliant of the disciples of Piet Mondrian, confessed, "My studies at school and at the Beaux Arts went hand-in-hand with Euclidean geometry. . . . The word 'space' especially excited my curiosity though I didn't know exactly why. Well, of course, it conformed to Euclidean geometry and all I had to do was submit." Vantongerloo speaks here for the tradition of twentieth-century art that extends from the Cubists through the architects of the Bauhaus and artists like Mondrian to the International Style and the Abstract Expressionism of the 1950s.

Even before the First World War, the geometric ideal had been challenged by F. T. Marinetti, the brilliant and enigmatic father of futurism. Marinetti argued that motion, speed, simultaneity, and process are the shaping elements of modern experience. To represent contemporary reality, art should depict objects in motion. Movies and dance are truly modern art forms because they move. Painting is inherently static, but it can suggest motion through distortion of the image and techniques like the simultaneity of Marcel Duchamp's famous *Nude Descending a Staircase*. Poetry can express the new aesthetic by speaking directly of machines and motion, and it can enact its subject by choppy diction, short lines, rapid and kaleidoscopic sequences of images and phrases, and passages in which several texts are understood as occurring at the same time.

The futurist ideal in architecture is dynamic rather than geometric. Here the coincidence of the futurist ideal with scientifc fact is especially obvious. Even though a building may pretend to be a static form—a perfect cube or cylinder—in reality it is constantly expanding and contracting. Heat lengthens its columns and beams; cold makes them shrink. The U.N. Secretariat and the twin towers of the World Trade Center in New York lean this way and that in the wind. Their parts pull against one another. No matter how Euclidean they may look, they sway constantly like a long-stemmed flower. In spite of its powerful aesthetic appeal, in one sense the John Hancock building is defensive rather than progressive. Its tapering profile and the huge diagonal braces that are expressed on its façade assert the determination of its architects to keep it rigid in spite of itself.

In the 1960s American architects began to experiment with dynamic forms. Eero Saarinen was no futurist, but his Dulles Airport Terminal in

Virginia is a brilliant expression of the dynamic impulse. It rejects rectangles in favor of curves, but the curves are not simple. They are created by balanced tensions—by roof cables straining against cantilevered buttresses—rather than by a system of columns and beams. The terminal is held up by the tendency of its walls to fall down. It is efficient because cables in tension are, pound for pound, far stronger than steel columns in compression. Its aesthetic is not only more dynamic but also more truthful—and perhaps more natural—than the geometries of the International Style.

Suspension bridges are inherently dynamic. Since there is no reason to hide the fact, they express their dynamic visibly. Their main support cables hang in natural curves created by gravity. They are catenaries punctuated by vertical cables from which the gently rising arch of the roadway is hung. They are in constant motion as they adjust to changes in temperature, wind direction and velocity, and load. They are gigantic mobiles even though their motions are usually invisible.

Because the mathematical qualities of their structure are fully visible, the beauty of the larger and more impressive suspension bridges has been widely appreciated. This is especially true of the Brooklyn Bridge, which may be for America what the Eiffel Tower is for France. Joseph Stella's many pictures of it form one of the most powerful sustained celebrations of technological dynamics in twentieth-century American art. The Brooklyn Bridge also inspired one of the most remarkable American poems of the twentieth century, Hart Crane's *The Bridge*, published in 1930. From the same room in which John Roebling watched his bridge being constructed Crane wrote:

> O harp and altar, of the fury fused,
> (How could mere toil align thy choiring strings!)
> Terrific threshold of the prophet's pledge,
> Prayer of pariah, and the lover's cry,—
>
> Again the traffic lights that skim thy swift
> Unfractioned idiom, immaculate sigh of stars,
> Beading thy path—condense eternity:
> And we have seen night lifted in thine arms.

Eero Saarinen, Dulles Airport, northern Virginia (1961). The form is organic rather than severely geometric. It is also dynamic, since the roof is supported by tension created by the outward and downward thrust of the vertical walls.

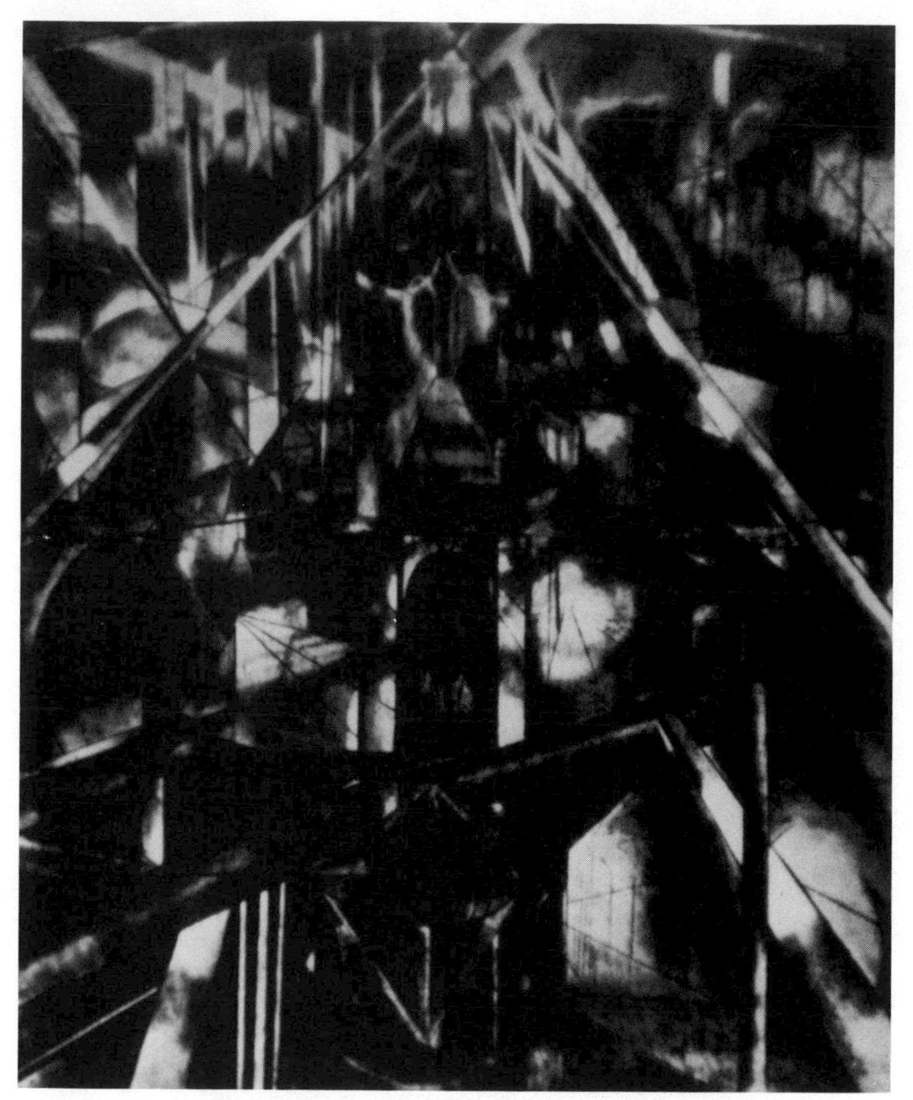

Joseph Stella, Brooklyn Bridge *(1917). Stella's artistic love affair with the Brooklyn Bridge parallels the fascination of the French artist Robert Delaunay with the Eiffel Tower.*

This is an extraordinary attempt to force language to express a dynamic aesthetic, and for anyone interested in the relations between culture and technology the poem will repay any number of attentive readings.

Its central quality is tension. Tension is expressed through the metaphor of harp strings. Crane imagines the great catenaries formed by the main cables between the New York and Brooklyn towers as the frame, and the support cables as the strings of a harp. Because the support cables are under enormous tension, they vibrate, creating a suprahuman music. The motif of tension is extended by images of things connected that are normally separated by great distances: the lights of the cars moving over the bridge and the lights of the stars circling above it; the curves of the main cables and the immense curve of the night sky. Everything is moving, vibrating, singing. The dynamism is reinforced by words suggesting motion—"traffic," "swift," "path," "skim," "lifted"—and by words that imply powerful, barely controlled energies—"fury," "prayer," "cry."

The poem also has problems, and they are as instructive as its successes.

In spite of the fact that Crane chose a great product of modern technology as his subject, he was no futurist, and he insists in a 1929 essay, "Modern Poetry," that machines have really not changed anything for the poet. The task of poetry is to absorb science and technology and "convert . . . experience into positive terms." A distinction is thus set up between the neutral values of science and the "positive terms"—meaning affirmative emotional responses—of art. Science and art do not conflict because they minister to different human needs. Crane notes that William Blake's "a tear is an intellectual thing" can be true artistically even though science would have a hard time explaining it.

Since Crane believes that science and art are by nature separate and committed to different methods, he also believes there can be continuity in modern art between past and present. The old human values are eternal. They survive even the most violent scientific revolutions. History is a means of salvation, of perspective in a world of flux.

The idea is familiar. It is associated with an attitude that is found everywhere in twentieth-century art but is perhaps more common in literature than in other fields. The trouble with it is that it turns out in Crane's case to be a formula for artistic failure. Marinetti felt it would always be a formula for artistic failure. Technological culture is a new departure, and to ignore that fact is to die artistically. For Marinetti the choice is between making art and making museum pieces: ". . . museums, libraries, and academies are, for artists, as damaging as the prolonged supervision by parents of certain young people drunk with their talent and their ambitious wills. When the future is barred to them, the admirable past may be a solace for the ills of the moribund, the sickly, the prisoner. . . . But we want no part of it, the past."

For one thing, Crane knew too much. What he knew is what one would expect a sensitive young man from Garrettsville, Ohio, to know—history. Ezra Pound, born in Hailey, Idaho, had the same problem and wrestled with the same limitations. Throughout Pound's career the impulse to "make it new" mixes exotically with the compulsion to stuff poems with history. Crane is not as obsessed as Pound. He is not out to prove in every poem that he can pass an oral examination in comparative literature. His problem is more like that of an art student eager to show he has mastered all the drawing exercises at the École des Beaux Arts.

Unfortunately, Crane *had* mastered the exercises, or (perhaps more accurately) the exercises mastered him. The dominant meter of *The Bridge* is iambic pentameter understood and written in the manner of Christopher Marlowe and Shakespeare. Beside the free verse of Pound and Eliot it often seems predictable and old-fashioned. Crane's vocabulary and diction are a mixture of influences in which traces of Milton and Shelley mix with recollections of Walt Whitman. The sense of decorum they reveal is Victorian. Lapses in tone, rhythm, and diction litter the poem like potholes in a highway. Polysyllables are equated with elevation and archaisms with powerful emotion. In the first eight lines we encounter "unfractioned idiom," "immaculate sigh," "choiring strings," "prayer of pariah," and "thine arms."

In spite of the fact that Roebling's bridge is a brilliant triumph of high technology, Crane's theme is traditional. It is a variant of Walt Whitman's grandiose effort in *Leaves of Grass* to see America whole. The theme eventually threatens to swallow up the Brooklyn Bridge, so that even as Crane tries to name it, it disappears into the poem, becoming a silence.

Or almost a silence, because *The Bridge* is not completely buried under history. The real burns at its center and is constantly forcing its way to the surface, exploding the syntax, overwhelming the meter, and creating a new language that expresses in sound the realities that old words cannot name:

> The nasal whine of power whips a new universe . . .
> Where sprouting pillars spoor the evening sky,
> Under the looming stacks of the gigantic power house

Stars prick the eyes with sharp ammoniac proverbs,
New verities, new inklings in the velvet hummed
Of dynamos where hearings leash is strummed . . .
Power's script—wound, bobbin-bound, refined—
Is stropped to the slap of belts on booming spools. . . .

In this passage Crane manages to break through the traditions that so often come between himself and his subject. Having written it, he must have recognized the deficiencies that are evident in so many other passages. The awareness accounts for the sense of bafflement and anguish that pervades his work. He wants desperately to name the world he has seen, but the artistic language he learned from books is all art. The opening lines of "Legend," the first poem in Crane's *White Buildings* (1926), are "As silent as a mirror is believed/Realities plunge in silence by." It is a brilliant and precise image for Crane's problem and the problem of language in general in modern culture.

A contrast. For years the most popular attraction at the Smithsonian Institution's National Air and Space Museum has been a film produced by Francis Thompson, with remarkable photography by Greg MacGillevray and James Freeman, entitled *To Fly*. The film is projected on a special screen fifty by seventy-five feet, and its sound comes from thirty loudspeakers. John Fialka, a journalist, described the film in the *Washington Star*: "The A4 jets of the Blue Angels ram through the air over the canyon in precise delta formation, wing tip a few inches from wing tip. It is man's symmetry, contrasted starkly with the splendors below. But man's work has its own beauty. The jets wheel upward and suddenly the screen is split into dozens of tiny frames. The lilting harmony of a Vivaldi-like score comes over the speakers."

Later, Fialka describes a sequence showing the launching of a Saturn rocket: "Witness the last Saturn flight as it begins in the film, balancing tentatively on the furies of its engines, yet reaching for a pinpoint in space. It is a summa of facts and figures, of f-stops and risks calculated by men who were, like Freeman, upbeat."

Fialka is reaching for the kind of poetry that Crane attempts in *The Bridge*. He contrasts dazzling and risky precision flying, which is also "man's symmetry," with the "splendors" of nature. A strong geometric form—a silver delta of A4 jets—is outlined by the rough landscape below it. As the planes veer upward, the art of cinema intensifies the effect. The movie screen dissolves into a shower of tiny images accompanied by a "Vivaldi-like" score.

In the second passage Fialka drops the pretense of narration. The report takes on the quality of live coverage. "Witness" is a direct command. It implies that reader and reporter are present at the scene. As the rocket begins its ascent, Fialka pauses to admire the photographer's command of the technology of "f-stops and risks." Risk and precise control are characteristic of technological art. They link the artist to the world of men who dare to guide Saturn rockets to "a pinpoint in space."

The beauty of jet flight and rockets is a modern and dynamic kind of beauty. *To Fly* flashes by in moving images separated by abrupt transitions in a way that would have delighted F. T. Marinetti. Fialka's prose is better at evoking technological culture than Crane's poetry because it is not crippled by history. The language is direct, simple, idiomatic, and free of the artificiality of meter. On the other hand, it remains language. Fialka knows it cannot do what a movie does. At critical moments in both of the passages quoted, language gives way to silence as Fialka turns from the scene itself to the technique of the photographer.

Crane and Fialka are both modern artists because both attempt to find languages adequate to the realities of modern life. *The Bridge* is a classic, but it is flawed by its inability to shake off the past. No one would call Fialka's description of *To Fly* a classic, but it is a lively and effective vignette of a typically modern experience.

14

MODERN VERSUS MODERNIST

Modern art is exactly what the term suggests. It is art that has found ways to name modern experience. As developments in the visual arts between 1900 and 1910 demonstrate, discovering these names required a radical break with the past. The break in the visual arts is illustrated by the explosion of schools and manifestos in the early twentieth century—Cubism, futurism, Expressionism, Constructivism, Dada, De Stijl, Vorticism, and a dozen more—and by the movement to techniques like abstraction and collage and simultaneity and the art of the readymade.

One of the first American writers to sense what it might be to be modern was Gertrude Stein. It is significant that she prepared at Radcliffe College for a career in psychology and studied with William James, and that she later spent three years at Johns Hopkins Medical School. It is also and obviously significant that she spent the decade between 1900 and 1910 in Paris, where she became friends with Matisse, Picasso, and most of the other painters who were creating modern art. Among the major pieces she wrote before 1914 were *Three Lives*; the sketches of Cézanne, Matisse, and Picasso published by Alfred Stieglitz in *Camera Work* in 1912; *Tender Buttons*; and the "Portrait of Mabel Dodge." These pieces trace a literary development so radical that many serious readers have still not caught up with it. In the closing years of the twentieth century these early pieces are still breathtaking and tantalizing suggestions of what might have been if America had been more receptive to the implications of modern, in contrast to modernist, literature.

Let us consider the issue of modern versus modernism more closely. Such sympathy as modern literature received in America was due largely to a group of painters and writers living in New York between 1910 and 1920. Walter Arensberg was the patron of the group, and Marcel Duchamp, who stayed with Arensberg during the war years, was an important link between European and American experimental artists. The Armory Show of 1913 played a central role in alerting Americans to the new developments in art, but it did little to stimulate interest in the small group of avant-garde writers associated with the Arensberg circle.

William Carlos Williams is representative of the effort of writers of this New York group to create a modern in contrast to a modernist literary art. He was familiar with futurism. Perhaps he was able to assimilate its lessons easily because he was not a student of literature or painting but a physician. Science was something immediate and familiar, and like Gertrude Stein, Williams was in an excellent position to appreciate just how dramatically science was changing traditional views of reality.

Wherever he learned what it was to be modern, the results appear early in his poetry and affect its deep structure. They are evident in the simple, standard American vocabulary, the colloquial diction, the rejection of meter, the abrupt, fragmented lines, the suppression of connectives, the dazzling shifts of perspective, and the colors and details that appear like fragments in a moving field of vision. Almost from the beginning, Williams writes poetry that is free of the pall of nineteenth-century literary conventions.

However, in spite of the nominal recognition which Gertrude Stein has received and the more genuine recognition accorded to Williams, a list of twentieth-century writers in the literary mainstream would be almost exclusively modernist. Along with Ezra Pound it would include T. S. Eliot, William Butler Yeats, James Joyce, D. H. Lawrence, Marcel Proust, François Mauriac, André Malraux, Rainer Maria Rilke, Thomas Mann, Ernest Hemingway, and William Faulkner. More names could readily be added to the list, but additions would merely make the bias more obvious, and it is obvious enough already. The writers included are all thoroughly familiar to readers and have been thoroughly assimilated by the academy. For a great many people interested in literary matters, they *are* twentieth-century literature.

Are they? In an interview in Bram Dijkstra's *A Recognizable Image*, Williams remarks, "It was the great period of Picasso's supremacy, of Braque, of Juan Gris, Matisse and some of the others. Do you find any inkling of that in what Pound was writing those days? . . . Picasso snubbed him. Gertrude Stein put him aside after a few words. . . . Briefly, Pound missed the major impact of his age. . . . Really, he can't learn and as a result has been left sadly in the rear."

This is a sensible remark, and it helps to clarify the difference between *modern* and *modernism*. Modern art recognizes a radical discontinuity between past and present and affirms the present. Related to this is the tendency of modern—in contrast to modernist—art to accept and celebrate technology, as in Marinetti's futurism and Bauhaus architectural theory and Hart Crane's decision—no matter how qualified—to make the Brooklyn Bridge the central symbol of his poem about America. For modern art, America is often a symbol of affirmation of the new in contrast to Europe, which remains chained to the past. Hence the impact of jazz in Europe during the 1920s and, two decades later, the delight of discovery that spills out of Mondrian's series of paintings titled *Broadway Boogie-Woogie*.

Modernist writers recognize that things have changed, but their typical response is nostalgia for the past in a world made unbearable by its loss. Often the loss is associated with technology and mass culture and—as shorthand for all of these—"America." America is said to mean automobiles and advertising and movies (and more recently, television) and lurid journalism and invasions of privacy and cheap ready-made clothing and the reducing of everything to the level of the beer-drinking masses, not to mention, in more recent times, fast foods, Richard Nixon and Jerry Falwell and Johnny Carson. The German poet Rainer Rilke sounds the true modernist note in a letter to his Polish translator in 1925: "From America empty, indifferent things are crowding over to us, sham things, *life-decoys.* . . . Animated things, things experienced by us, and that know us, are on the decline and cannot be replaced anymore. *We are perhaps the last still to have known such things.*"

T. S. Eliot's *The Waste Land* is a powerful and characteristic expression of the modernist point of view. It is a lament for everything that has been lost from modern life. It draws its basic symbolism from the story of the search for the Holy Grail, and its recurrent theme is the contrast between the life-sustaining richness of the past and the drabness of the twentieth century. A similar theme runs through the poetry of W. B. Yeats, who began his poetic career with languid evocations of Irish mythology and was fond of contrasting the values of an imaginary heroic Irish past with those of a present corrupted by machines and commerce. In much modernist writing, dislike of technology emerges as pastoralism, which combines nostalgia for an Edenic past with the idea that those who live close to nature are somehow nobler than those who do not. The pastoral note sounded with specifically American historical overtones in the nostalgia of the so-called Fugitive-Agrarian writers for the values of the agrarian South, as, for example, in their 1930 manifesto, *I'll Take My Stand.* Ernest Hemingway favored poor but wise peasants and honorable sportsmen to Jeffersonian farmers, and the two types merge in the quintessential Hemingway story *The Old Man and the Sea.*

Evidently, one problem of modernism is that even when it uses radical techniques, it tends to celebrate an idealized otherness rather than to struggle in the manner of Mallarmé and Hart Crane to come to terms, if only grudgingly, with the here and now. A second problem is that nostalgia can easily become reaction. The novelist François Mauriac supported a quasi-medieval Catholicism that claimed the right to shape lay opinion on political and social as well as religious matters. T. S. Eliot repeatedly cited Dante as the defining example of what the poet can achieve in a supportive culture, but he settled for being, as he put it, an Anglo-Catholic and a Royalist.

In *The Idea of a Christian Society*, written in 1939 on the eve of the Second World War, Eliot went further. He confessed that he had no idea what democracy means and suggested that "defenders of the totalitarian [i.e., Nazi] system can make out a plausible case for maintaining that what we have is not democracy, but financial oligarchy." Ezra Pound broadcast propaganda for Mussolini during the Second World War, and the novelist Louis-Ferdinand Céline collaborated with the Germans during the occupation of France. Anti-Semitism runs like an ugly thread through the writing of many modernists, although one can obviously be modernist without being racist, and—as the case of Marinetti shows—being modern is no infallible protection against being fascist.

At its best, modernist art seems to be art that honors the past because tradition is a protection against excess, in contrast to modern art, which is always on the point of declaring that history is a burden. The term *modernist* calls attention to the use of radical techniques to convey this essentially conservative theme. In this respect, as in others, Eliot's *The Waste Land* is a defining model. The poem looks radical. In certain respects it *is* radical since it draws on Ezra Pound's formulas for creating a new poetic language through colloquial diction, avoidance of obviously "poetic" metaphors and rhythms, and communication through small, energy-charged packets of statement or image rather than sustained narration.

At the same time, the poem is a long lament for the loss of an earlier purity. Its standard means of evoking history is a literary figure—allusion. Marinetti complained that too much art resembles a museum display. Eliot, conversely, made a poem that is a kind of literary museum in which the allusions are the exhibits. To make sure the reader does not miss the allusions, Eliot identifies them in footnotes. Later commentaries on *The Waste Land* have extended the number of allusions far beyond those that Eliot chose to put in glass cases. Even without the additions, however, the allusions span the entire history of Western literature from Ezekiel to Dante to Shakespeare to Baudelaire to Wagner to Sir James Frazer. The ghosts Eliot summoned have a single message: judged by the standards of the past the modern age is corrupt and degenerate.

Eliot's best statement of the modernist credo is the essay "Tradition and the Individual Talent." To those who insist first and foremost on originality Eliot replies that the most important talent is historical:

> . . . the historical sense . . . we may call nearly indispensable to anyone who would continue to be a poet beyond his twenty-fifth year; and the historical sense involves a perception, not only of the pastness of the past, but of its presence; the historical sense compels a man to write not merely with his own generation in his bones, but with a feeling that the whole of the literature of Europe from Homer and within it the whole of the literature of his own country has a simultaneous existence and composes a simultaneous order.

Most readers agree that *The Waste Land* enacts history in exactly this way. They are wrong. Eliot's use of history does not lead to a sense of simultaneous order—or any other order, for that matter. *The Waste Land* sputters out in a pastiche of quotations that seem to add up to madness or nonsense:

> *Le Prince d'Aquitaine à la tour abolie*
> These fragments I have shored against my ruins
> Why then Ile fit you. Hieronymo's mad againe.
> Datta. Dayadhvam. Damyata.
> Shantih shantih shantih

This is neither a successful retreat from the present into a dream of pastoral fulfillment nor a vision of the present redeemed by the values of the past. It is an expression of total bafflement; a kind of silence, you might say. It enacts a moment in which history is caught in the act of disappearing even as it struggles to assert its presentness. As such, it is a comment on much other modernist literature. Pastoralism doesn't work, whether the retreat is understood as a trout stream far from the evils of the city or a peasant village in Sligo or the medieval Church or an authoritarian state.

15

RUNNING FENCE

The sculptor Christo, surnamed Javacheff, received national publicity in 1971 when he announced plans for *Running Fence*, a post, wire, and fabric sculpture extending twenty-four and a half miles across the hills of Marin and Sonoma counties in California. It was an odd structure. There was no precedent for it—it had no history. A fence is supposed to keep something in or out, but Christo's fence simply "ran" (see illustration 4 in color section).

Although highway departments and power companies tear up tens of thousands of acres every year to build highways and string high tension lines, their activities go unremarked because they are considered useful. Christo's *Running Fence* was puny in comparison to highways and high tension lines, but it was noticed because, like the Eiffel Tower, it was useless.

In fact, its uselessness was what led to its notoriety. Since it ran across private and public lands, Christo had to seek permission from the ranchers whose lands he would cross and from the zoning boards of Marin and Sonoma counties. The application stirred up widespread opposition which increased as successive applications were turned down, rewritten, and resubmitted to the zoning boards.

Some felt the fence was absurd. Some felt it was ugly and insulting. Many ranchers were initially afraid that it would attract mobs who would trample their fields and frighten their cattle. Environmentalists protested that it would rape the ecology of northern California. Each protest brought out defenders. Eventually, Christo's allies won the day. In the process they accomplished what Christo had anticipated from the beginning. They converted *Running Fence* from a minor oddity deserving at the most a couple of paragraphs in a local paper to an event with international visibility. *Running Fence* became an announcement to the world that technology had created a new kind of aesthetic—the aesthetic of linear structure. It is an aesthetic created by the imposition of lines on an undefined, often extremely irregular surface. As such, it has clear affinities to the aesthetic of the linear abstractions of artists like Mondrian and Vantongerloo, which is to say that it is transhistorical.

Christo argues that art must be measured by its effect on the shape of a community. By this criterion *Running Fence* was enormously successful. It first polarized and then reunified the community in which it was created. From Christo's point of view it might be correct to call the new shape of the community the permanent artwork, since it persisted after *Running Fence* was dismantled. On the other hand, *Running Fence* itself was the thing that caused the reshaping of the community because it was the thing the community was forced to see. Without *Running Fence* the community would have seen nothing and therefore would have done nothing. In this sense it is possible to call *Running Fence* the artwork.

The most persuasive claim of *Running Fence* to being art is that it is an imitation of reality. The aesthetic of linear structure is characteristic of the twentieth century, which throws such structures across the modern landscape in great profusion: railroads, telegraph and telephone lines, high tension lines, pipelines, highways. Such structures are everywhere. They register on our consciousness and silently shape it. Christo's fence is, in effect, a name for this effect.

1. James Audubon, *Ivory-Billed Woodpeckers*. From *Birds of North America* (1827–1838).

2. Infrared Landsat photograph of Cape Hatteras (NASA, 1978). Nature seen from a great distance and with selective screening of data.

3. Benoit Mandelbrot, *Fractal Image* (1988), showing the brilliantly aesthetic qualities of fractal images in color.

4. (ABOVE) Christo, *Running Fence* (1972). The fence has no functional use; it must therefore be seen in aesthetic terms as a frame for the natural landscape and an assertion of linearity in contrast to the irregular natural contours of the hills through which it cuts.

5. (RIGHT) *Mirror Building* — typical of the surreal imagery created by the facades of buildings covered by reflective glass. Photograph by Stavros Moschopoulos/Image Ray, 1988.

6. Melvin Prueitt, *Gold Wing* (1981) — advanced computer art that calls attention to its mathematical ancestry.

7. David Em, *Bill* (1982). One of the classics of computer-

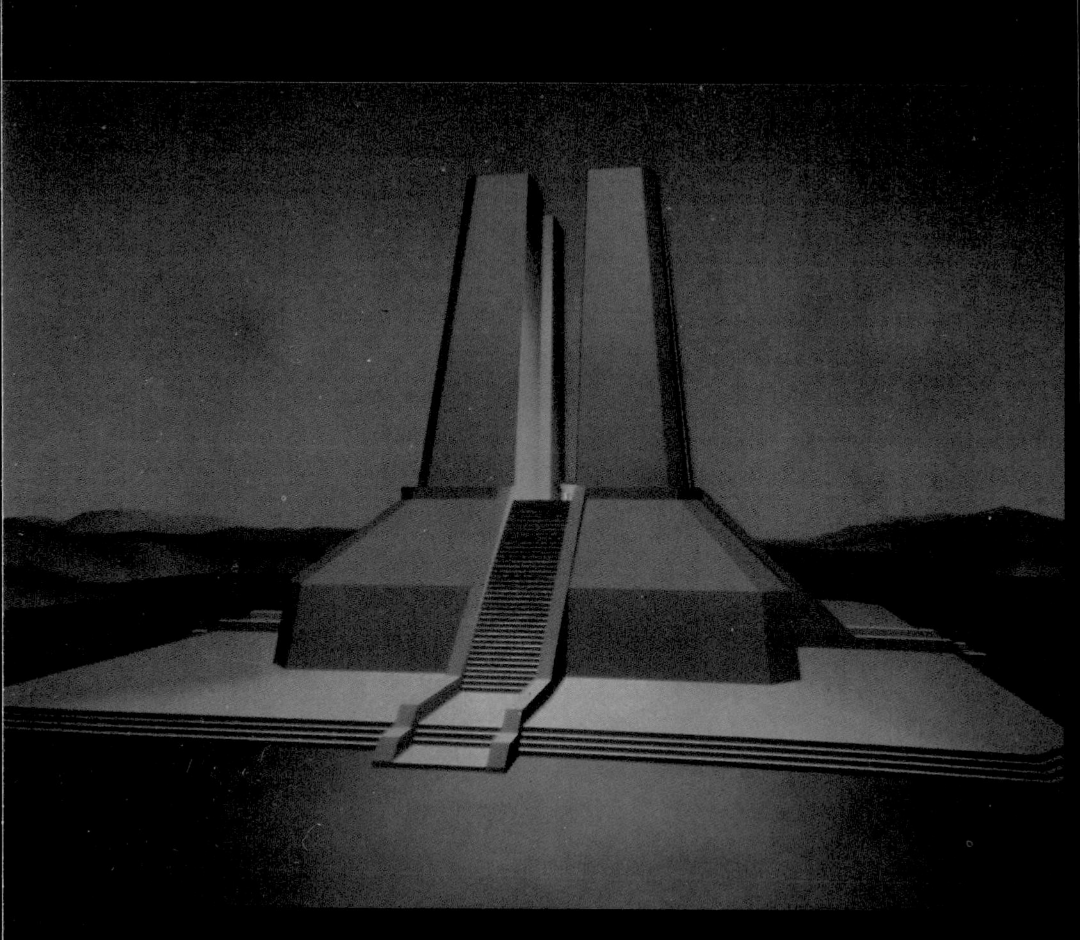

8. Computer-generated temple — a fusion of high tech and mystery (Digital Productions, 1983).

Considered in relation to *Running Fence*, the dual-lane highways of the American interstate system can be recognized as the most majestic linear sculptures created in human history, far surpassing in size, conception, and engineering the Great Wall of China. Seen from the air they create grand reticulations, making the flat places of the earth look like huge Mondrian paintings. Seen from the ground they segment and order the landscape. Perhaps they should be understood as picture frames because they make pictures out of what would otherwise be disordered and undefined.

Pioneers approaching the Sierra Nevada mountains in the 1850s must have regarded them with awe and apprehension. If the season was late, perhaps stories of the ill-fated Donner party—its wagons trapped in bottomless snow drifts, the passengers reduced to cannibalism and finally to frozen corpses—would circulate among the wagons.

Today's driver moving toward the mountains west of Salt Lake City through a ruddy late-fall sunset feels delighted and soothed. The snow-covered peaks glisten, turn pink, and then fade into the soft, powdery blue. The gentle grades, sweeping curves, and graceful bridges of the Interstate assert the triumph of man over nature.

Nature is elevated by this triumph. It becomes a form of art: scenery. The bracing effect of a human reference point on the vast disorder of nature is the subject of Wallace Stevens's "Anecdote of the Jar":

> I placed a jar in Tennessee,
> And round it was, upon a hill.
> It made the slovenly wilderness
> Surround that hill.
>
> The wilderness rose up to it,
> And sprawled again, no longer wild.
> The jar was round upon the ground
> And tall and of a port in air.
>
> It took dominion everywhere.
> The jar was gray and bare.
> It did not give of bird or bush,
> Like nothing else in Tennessee.

We take dual-lane highways for granted. We have grown up with them. They seem "natural," which is to say that they seem part of the way things are. However, if we were travelers from a roadless civilization looking at them for the first time, we would be as awed by their boldness as the Goths were when they first entered the cities of Gaul.

In antiquity, temple architecture and religious sculpture were complementary. Each was created with the other in mind. A temple was planned for the statue of a god, and a large-scale statue of a god was usually created for a specific temple. A similar relationship links modern dual-lane highways with the most popular three-dimensional art form ever created, the thin-steel sculpture known as the automobile.

Considered as art, automobiles give rise to a flagrant kind of hypocrisy. We read about them constantly, agonize over different makes and models, pay exorbitant prices for them, and polish and manicure them, often with greater care than we lavish on our children. At the same time, we pretend they are utilitarian objects whose appeal, if they have any beyond utility, is psychological: they are status symbols, coming-of-age symbols, symbols of virility, symbols of independence.

They may, of course, be all of these things, but they are also thin-steel sculptures expressing the aesthetic of speed. F. T. Marinetti, who called himself "the caffeine of Europe," describes this aesthetic in his 1909 *Manifesto*:

> We assert that the magnificence of the world has been enriched by a new beauty, the beauty of speed. A racing car with its bonnet draped in enormous pipes like fire-spitting serpents . . . a roaring car that goes like a machine gun, is more beautiful than the *Winged Victory* of Samothrace.

Carl Sandburg evokes the same aesthetic in "Portrait of a Motorcar." The poem shows that language can be both immediately accessible and convincingly dynamic:

> It's a lean car . . . a long-legged dog of a car . . . a gray-ghost eagle car.
> The feet of it eat the dirt of a road . . . the wings of it eat the hills.

Danny the driver dreams of it when he sees women in red skirts and red sox in his sleep.
It is in Danny's life and runs in the blood of him . . . a lean gray-ghost car.

The automobile was born in 1885. For fifty years designers covered automobiles with historical and symbolical decoration. Naked ladies and flying birds perched on their radiator caps. Flowers sprouted from vases in their passenger compartments. Early limousines separated the chauffeur, who rode in an open compartment, from the passengers, who were snugly isolated in a luxurious waterproof box. The arrangement was taken over from the design of coaches: since the coachman held the reins, he had to ride outside. It was retained in early limousines in spite of the fact that failure to protect the chauffeur dramatically reduced the safety of the passengers.

Between 1950 and 1965 General Motors put tail fins on its cars to intimate that they were rockets. This is a modernist strategy: the concept of speed expressed by an extraneous symbol rather than the thing itself.

But the logic of automotive design has nothing to do with decorated radiator caps or flowerpots or tail fins. It was first embodied fully in a production model car by Carl Breer in the Chrysler Airflow of 1934. The most obvious feature of the Airflow is streamlining. The term was invented

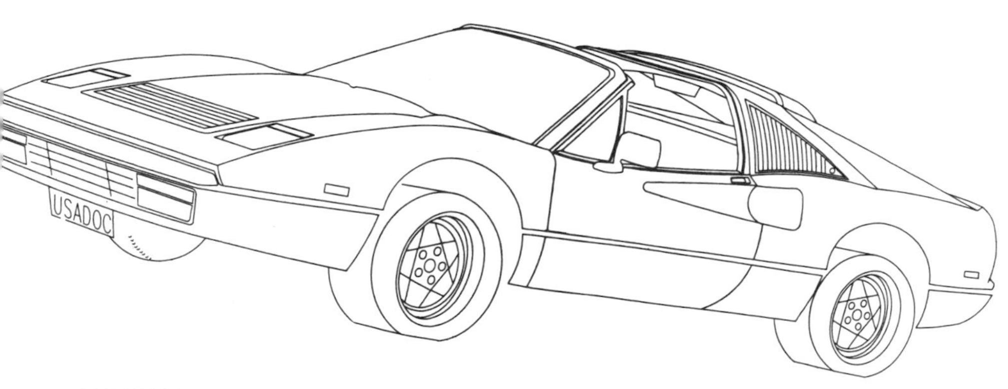

CAD/CAM sports car (1988).

by D'Arcy Thompson in *On Growth and Form* to describe the curvature imposed by water on the body of a fish. It is not a symbol of speed like a tail fin; it is the technological condition for efficient high-speed travel through a fluid.

The Airflow embodied several other design features reflecting the logic of technology. It used a unified steel frame for the body, which greatly reduced rattles, and its skin was stamped from a single piece of stressed steel, which not only reduced rattles but lowered production costs. Weight distribution was calculated in relation to the distance between axles to provide maximum passenger comfort.

The Airflow was to automobiles what the Crystal Palace was to buildings. The car expressed speed in its design. It was a symbol of what it was. It was also an unmitigated financial disaster. It departed too radically from the boxy design of its predecessors. It was modern, not modernist, and the public wanted modernism—cars whose design alluded to tradition. The disappearance of the Airflow from the showrooms did not, however, signal its disappearance as an idea. The Airflow expressed technological imperatives. Such imperatives can be ignored, but they cannot be repressed permanently. Ferdinand Porsche adopted Breer's concepts when he designed the original Volkswagen. Since the Volkswagen, streamlining has become commonplace. It is now considered beautiful, and that fact is a lesson in the way that technology changes aesthetic perception even though it is initially rejected.

Sheldon Cheney's *Primer of Modern Art* is a classic of popularization. By 1945 it had gone through eleven editions and had served to introduce a whole generation of Americans to its subject. For Cheney the automobile exemplifies the entire modern aesthetic:

> While we deplore the lack of inventiveness and the reliance on imitative, run-out ornament in our furniture-making, our hardware and our chinaware, we are prone to overlook a beauty that is wholly and typically modern in our everyday machinery. The ordinary hand phone has its values in the directness with which it is designed for its purpose, and in the simplicity and the relationships of its lines and volumes. The machinery in the power-house has a potent line-and-form fascination that anyone alert to art must feel. But most common in experience today is the aesthetic value of the motor-car. . . . The

sheer volume-design of the automobile, its dependence on stream lines and expressive mass instead of ornament . . . and its absolute sense of fleetness, are qualities that, within the field of the arts of use, speak art-sense, and qualities to which we respond instinctively.

16

DISAPPEARING THROUGH THE SKYLIGHT

Science is committed to the universal. A sign of this is that the more successful a science becomes, the broader the agreement about its basic concepts: there is not a separate Chinese or American or Soviet thermodynamics, for example; there is simply thermodynamics. For several decades of the twentieth century there was a Western and a Soviet genetics, the latter associated with Lysenko's theory that environmental stress can produce genetic mutations. Today Lysenko's theory is discredited, and there is now only one genetics.

As the corollary of science, technology also exhibits the universalizing tendency. This is why the spread of technology makes the world look ever more homogeneous. Architectural styles, dress styles, musical styles—even eating styles—tend increasingly to be world styles. The world looks more homogeneous because it *is* more homogeneous. Children who grow up in this world therefore experience it as a sameness rather than a diversity, and because their identities are shaped by this sameness, their sense of differences among cultures and individuals diminishes. As buildings become more alike, the people who inhabit the buildings become more alike. The result is described precisely in a phrase that is already familiar: the disappearance of history.

The automobile illustrates the point with great clarity. A technological innovation like streamlining or all-welded body construction may be rejected initially, but if it is important to the efficiency or economics of automobiles, it will reappear in different ways until it is not only accepted but universally regarded as an asset. Today's automobile is no longer unique to a given company or even to a given national culture. Its basic features are found, with variations, in automobiles in general, no matter who makes them.

A few years ago the Ford Motor Company came up with the Fiesta, which it called the "World Car." Advertisements showed it surrounded by the flags of all nations. Ford explained that the cylinder block was made in England, the carburetor in Ireland, the transmission in France, the wheels in Belgium, and so forth.

The Fiesta appears to have sunk without a trace. But the idea of a world car was inevitable. It was the automotive equivalent of the International Style. Ten years after the Fiesta, all of the large automakers were international. Americans had plants in Europe, Asia, and South America, and Europeans and Japanese had plants in America and South America, and in the Soviet Union (Fiat workers refreshed themselves with Pepsi-Cola). In the fullness of time international automakers will have plants in Egypt and India and the People's Republic of China.

As in architecture, so in automaking. In a given cost range, the same technology tends to produce the same solutions. The visual evidence for this is as obvious for cars as for buildings. Today, if you choose models in the same price range, you will be hard put at 500 paces to tell one make from another. In other words, the specifically American traits that lingered in American automobiles in the 1960s—traits that linked American cars to American history—are disappearing. Even the Volkswagen

Beetle has disappeared and has taken with it the visible evidence of the history of streamlining that extends from D'Arcy Thompson to Carl Breer to Ferdinand Porsche.

If man creates machines, machines in turn shape their creators. As the automobile is universalized, it universalizes those who use it. Like the World Car he drives, modern man is becoming universal. No longer quite an individual, no longer quite the product of a unique geography and culture, he moves from one climate-controlled shopping mall to another, from one airport to the next, from one Holiday Inn to its successor three hundred miles down the road; but somehow his location never changes. He is cosmopolitan. The price he pays is that he no longer has a home in the traditional sense of the word. The benefit is that he begins to suspect home in the traditional sense is another name for limitations, and that home in the modern sense is everywhere and always surrounded by neighbors.

The universalizing imperative of technology is irresistible. Barring the catastrophe of nuclear war, it will continue to shape both modern culture and the consciousness of those who inhabit that culture.

This brings us to art and history again. Reminiscing on the early work of Francis Picabia and Marcel Duchamp, Madame Gabrielle Buffet-Picabia wrote of the discovery of the machine aesthetic in 1949: "I remember a time . . . when every artist thought he owed it to himself to turn his back on the Eiffel Tower, as a protest against the architectural blasphemy with which it filled the sky. . . . The discovery and rehabilitation of . . . machines soon generated propositions which evaded all tradition, above all, a mobile, extra human plasticity which was absolutely new. . . ."

Art is, in one definition, simply an effort to name the real world. Are machines "the real world" or only its surface? Is the real world that easy to find? Science has shown the insubstantiality of the world. It has thus undermined an article of faith: the thingliness of things. At the same time, it has produced images of orders of reality underlying the thingliness of things. Are images of cells or of molecules or of galaxies more or less real than images of machines? Science has also produced images that are pure artifacts. Are images of self-squared dragons more or less real than images of molecules?

The skepticism of modern science about the thingliness of things implies a new appreciation of the humanity of art entirely consistent with Kandinsky's observation in *On the Spiritual in Art* that beautiful art "springs from inner need, which springs from the soul." Modern art opens on a world whose reality is not "out there" in nature defined as things seen from a middle distance but "in here" in the soul or the mind. It is a world radically emptied of history because it is a form of perception rather than a content.

The disappearance of history is thus a liberation—what Madame Buffet-Picabia refers to as the discovery of "a mobile extra-human plasticity which [is] absolutely new." Like science, modern art often expresses this feeling of liberation through play—in painting in the playfulness of Picasso and Joan Miró and in poetry in the nonsense of Dada and the mock heroics of a poem like Wallace Stevens's "The Comedian as the Letter C."

The playfulness of the modern aesthetic is, finally, its most striking—and also its most serious and, by corollary, its most disturbing—feature. The playfulness imitates the playfulness of science that produces game theory and virtual particles and black holes and that, by introducing human growth genes into cows, forces students of ethics to reexamine the definition of cannibalism. The importance of play in the modern aesthetic should not come as a surprise. It is announced in every city in the developed world by the fantastic and playful buildings of postmodernism and neomodernism and by the fantastic juxtapositions of architectural styles that typify collage city and urban adhocism.

Today modern culture includes the geometries of the International Style, the fantasies of façadism, and the gamesmanship of theme parks and museum villages. It pretends at times to be static but it is really dynamic. Its buildings move and sway and reflect dreamy visions of everything that is going on around them. It surrounds its citizens with the linear sculpture of pipelines and interstate highways and high-tension lines and the delicate virtuosities of the surfaces of the Chrysler Airflow and the Boeing 747 and the lacy weavings of circuits etched on silicon, as well as with the brutal assertiveness of oil tankers and bulldozers and the Tinkertoy complications of trusses and geodesic domes and lunar landers. It abounds in images and sounds and values utterly different from those of the world of natural things seen from a middle distance.

It is a human world, but one that is human in ways no one expected.

Disappearing bank (San Francisco Bank—*photograph by James Wilson, 1987*).

The image it reveals is not the worn and battered face that stares from Leonardo's self-portrait, much less the one that stares, bleary and uninspired, every morning from the bathroom mirror. These are the faces of history. It is, rather, the image of an eternally playful and eternally youthful power that makes order whether order is there or not and that having made one order is quite capable of putting it aside and creating an entirely different one the way a child might build one structure from a set of blocks and then without malice and purely in the spirit of play demolish it and begin again. It is an image of the power that made humanity possible in the first place.

The banks of the nineteenth century tended to be neoclassic structures of marble or granite faced with ponderous rows of columns. They made a statement: "We are solid. We are permanent. We are as reliable as history. Your money is safe in our vaults."

Today's banks are airy structures of steel and glass, or they are storefronts with slot-machinelike terminals, or trailers parked on the lots of suburban shopping malls.

The vaults have been replaced by magnetic tapes. In a computer, money is sequences of digital signals endlessly recorded, erased, processed, and reprocessed, and endlessly modified by other computers. The statement of modern banks is "We are abstract like art and almost invisible like the Crystal Palace. If we exist at all, we exist as an airy medium in which your transactions are completed and your wealth increased."

That, perhaps, establishes the logical limit of the modern aesthetic. If so, the limit is a long way ahead, but it can be made out, just barely, through the haze over the road. As surely as nature is being swallowed up by the mind, the banks, you might say, are disappearing through their own skylights.

Robert Indiana, LOVE. *Photograph by Stavros Moschopoulos/Image Ray.*

PART III

THE POETRY OF NOTHING; OR, THE DISAPPEARANCE OF LANGUAGE

When the dogs cross the air in a diamond
like the ideas and the appendix of the meanings
show the hour of the awakening program

—TRISTAN TZARA,
"*To Make a Poem*"

17

THROWING DICE

Modern painting emerged with stunning suddenness in Western culture between 1900 and 1910. A good symbolic date for the moment of emergence is Picasso's *Les Demoiselles d'Avignon,* completed in 1907 and generally considered the first unambiguous statement that Cubism had arrived. If art expresses states of consciousness, the art that burst forth in such amazing profusion in the first decade of the twentieth century expresses a new state of consciousness that seems to have communicated itself through some mysterious spiritual medium even before it was explicitly defined.

When the artists cited influences outside of the world of painting that affected their work they ranged across the whole landscape of the culture: primitive African and Iberian art, disgust with Philistinism, hatred of class oppression, "the fourth dimension," relativity, Freudian psychology, technology, and more. All of these influences were undoubtedly operative. They were agents of cultural change as well as symptoms, and they intensified the awareness of artists of the need for complementary changes in visual art.

The fact that science is among the influences cited to explain the emergence of modern art does not mean there are simple correlations between modern science and modern art. The chief influences on the new art came from the world of art itself, just as the chief influences on the new physics came from physics. There were, however, significant common tendencies. In moving from representation toward abstraction, modern art rejected a vocabulary of representation that had been used by Western painting since the Renaissance and that had been generally equated by artists and viewers alike with "meaning" in the sense that it related the image presented to historical tradition and to the personality of the artist. In theory, a perfectly abstract painting can be as free of tradition and personality as a triangle or a randomly curving line. It is thus liberated from the need to "mean" in the old way.

The corollary of abstraction in art is universality. The meaning of a painting like *The Rest on the Flight to Egypt* is immediately apparent to a viewer who knows the New Testament. But it is opaque to a viewer who does not know the New Testament. Chinese calligraphy depends on the artful matching of brushwork and text. A Chinese poem will be opaque to a viewer who cannot read the text, even though the viewer may admire the graceful calligraphy. Conversely, an abstract painting is transparent. It should appeal more or less equally—at least in theory—to viewers from widely differing social backgrounds and cultures.

There are many virtues in transparency. Since abstract forms come from the spirit rather than nature, abstract art is deeply and uniquely human, a vehicle for what Kandinsky appropriately called "the spiritual" in art. And since abstract art imitates the forms and categories of the spiritual life of people in general rather than the traditions of this or that religion or nation-state, it complements the ideal of world unity that has been prominent in so many other areas of twentieth-century thought.

Abstraction and universality are also qualities associated with science. Werner Heisenberg makes just this point in his essay "The Tendency to Abstraction in Modern Art and Science." We live, he observes, in an era of world culture. Art must depict this situation. In doing so, it must move in the direction science has taken in the description of nature:

> The young man no longer sees his life merely in relation to the tradition, the country, the culture in which he has grown up but relates it to the whole world. . . . The tendency. . . answers to the tendency in science to regard the whole of nature as a unity. . . . The realization of this program has pushed the sciences on to ever higher levels of abstraction, and . . . the relation of our life to the whole spiritual and social structure of the earth will also be capable of artistic presentation only if we are ready to enter into regions more remote from life.

The new consciousness that emerged in art between 1900 and 1910 emerged at about the same time in literature. A good symbolic date for its appearance is 1897, the publication date of Stéphane Mallarmé's "Un Coup de dés"—"A Throw of the Dice."

In one sense, the poem is the culmination of a theme present in many of Mallarmé's earlier poems: "the difficulty of writing." In a deeper sense, however, it is not only a new poem but a new kind of language. Both its physical appearance and its unorthodox syntax and use of words announce that it breaks as decisively with traditional poetry as Picasso's *Les Demoiselles d'Avignon* breaks with what preceded it in painting.

The theme of the difficulty of writing runs deep in Mallarmé's work for a very simple reason—language changes slowly but culture is changing rapidly. Language in its traditional forms therefore becomes less and less complementary to the world it is supposed to represent. The poet, who is supposed to represent reality in language, responds to the exhaustion of language with bafflement and frustration. The simplest response is silence. If language cannot represent the world, why write? To be silent, however, is to give up, a form of despair.

Another possible response, which is the response of "Un Coup de dés," is to invent a way past silence. Mallarmé's poem enacts the process of invention, and its language and the strategies used to deploy that language *are* the invention. It is thus a kind of epic recounting the poet's response in language to the absence of a language able to name reality. One of its dominant images—evident in the passage quoted below—is that of a voyage. Another is flight—the attempt to reach beyond the earth—and falling back. The world within which these efforts are made is ruled by chance and ordered only by arbitrary acts of the mind. Each act is like a throw of the dice—hence Mallarmé's title. Mallarmé's achievement is as brilliant in its way as Picasso's, and his poem is a dictionary of techniques and motifs of twentieth-century experimental poetry.

"Un Coup de dés" announces its newness in its physical appearance. The poem was first printed in the review *Cosmopolis* a year before Mallarmé's death in 1898. This version is already typographically unorthodox. Not until 1913, however, when the poem was reprinted in *La Nouvelle Revue Française*, were Mallarmé's intentions regarding typography fully honored. This latter form is followed in the English translation of pages 2 and 3 below:

BE
that
the ABYSS

whitened
slack
maddened
on a slope
slides desperately
a wing
its own
al

ready fallen because the flight was badly planned
and covering the surges
just shaving the billows

deep within resumes

the shade buried in the deep by this other sail
to the point of adjusting
to the spread
its gaping deep so much that the shell
of a ship

pitched from side to side . . .

The conventional printed page is a rectangular block of type surrounded by white margins. Poems are more flexible, but they normally appear as neatly arranged stanzas or blocks of lines, and they, too, are centered on the page and surrounded by white margins. Such a page makes a simple and unambiguous statement. The paper is the ground. It has no message and is significant only as a neutral medium on which type is printed. It is white, which is actually a mixture of all colors but is normally perceived as no color—a metaphor for nothing. The type, conversely, is the figure. It is black, a metaphor for hardness, impenetrability, the real. It contains the message, which is presented in a series of pulses called lines. In prose the lines are all the same length; in poetry they are usually short and vary in length in terms of formulas like "decasyllable" and

SOIT
que

l'Abîme

blanchi
étale
furieux
sous une inclinaison
plane désespérément

d'aile

la sienne
par

avance retombée d'un mal à dresser le vol
et couvrant les jaillissements
coupant au ras les bonds

très à l'intérieur résume

[alternative
l'ombre enfouie dans la profondeur par cette voile

jusqu'adapter
à l'envergure

sa béante profondeur en tant que la coque

d'un bâtiment

penché de l'un ou l'autre bord

"octosyllable." Typefaces for both prose and poetry are usually continuous, varying only through occasional capitalized words and, in some works, movements from roman to italic type and back again.

Mallarmé's page is black-on-white, but otherwise it departs radically from the conventions. The words move raggedly forward. In the passage quoted, the lines on the left slide visually downward in accord with the movement they describe. A few lines extend from one margin to another as in prose, but more frequently the lines are fragments—a word or a phrase. There is no formula for the length of the fragments. Moreover, the fragments do not back up neatly against a left margin but appear anywhere between the left and right limits. The typography also violates convention. Some words are in small type, some in large, some in capitals, some in lower case, some in roman, some in boldface, some in italics. Words and phrases are sometimes far left, sometimes far right, sometimes centered, sometimes written one after the other, sometimes arranged stairwise in descending order, sometimes separated by one or more standard line spaces. The blank page is the nothingness—the silence—on which the words enact their epic journey, which is both a quest for meaning and a creation of meaning. Because of this, the standard relationship between figure and ground is displaced. The white space is expressive—part of the message rather than a neutral field for the type.

Mallarmé has not thrown away the dictionary. His words retain their standard meanings, but his grammar and syntax are fragmented. Each syntactic fragment tends to occupy a separate line. The fragments create moods and fantasies as suggested by their dominant words, but the words—and the moods and fantasies—change rapidly. They are hardly sensed before they dissolve. Like the images in a kaleidoscope, they take shape in the flux, register, and give way to new images, like glimpses of a world illuminated by a strobe light. Phrases begin with brave assurance and then stumble. The words needed to complete their meanings are separated from them by chasms of whiteness. Is the poem a collection of fragments? Sometimes it seems to be, but sometimes—as in the passage quoted—it sustains itself for several lines. Sometimes the words crawl along the page, their agonizingly difficult motion a representation of the crushing force of the silence against which they struggle. Sometimes they rush forward as though the resistance of the silence had momentarily slackened.

In editions that follow Mallarmé's printing instructions, the single page gives way to units of two pages. These units are also part of the expression. The two-page arrangement makes use of the barrier that normally separates pages at the center. Crossing the barrier becomes a significant event. In the passage quoted the reader is forced across by the syntax—"al . . . ready" (*par . . . avance*). Should the reader always read all of one page and then the next, as in a conventional book, or across the two-page spread from far left to far right? Never? Sometimes? Both? The poem is indeterminate. The reader often has alternatives and in choosing composes the poem as it goes along; he may compose a different poem on each reading. In this sense the poem at least hints at the idea of interactive art.

Near the end of the poem there is a reference to a constellation: Ursa Major, the constellation that points to the North Star, a symbol of direction. Is there a way out? No, because nothing stands still. As the words of the poem remind us, there is only "*veillant, doutant, roulant*"—watching, doubting, rolling. The star is a pinpoint of light, a flicker. It identifies a place to pause, not a direction. After every pause there will be another throw of the dice and the process of discovering a direction will begin again.

The subject of the poem is randomness. The world has no meaning, but the mind endlessly imposes order on it by arbitrary acts. Meaning is created by an act of will: the placing of one word after another. The progression of the words and phrases is the poet's victory over the absence of language. The image of a voyage that flickers through the poem is a memory of ancient epic, which celebrated the triumph of human will in a hostile universe. The great epic voyages led home or to the founding of new states, or they were voyages to new worlds. In Mallarmé's random universe there is no reason to choose one direction rather than another.

The constellation of Ursa Major is infinitely distant. It is a myth. Primitive societies inscribed such myths on the random dust of the stars. They imagined patterns and called them Ursa Major and Orion and Sagittarius. The act of naming patterns that do not exist and thus of bringing them into existence is a metaphor for making poems. Beginning with the title of the poem, the phrases in capitals march through the pages like a continuing voice: "A THROW OF THE DICE . . . NEVER . . . WILL ELIMINATE . . . CHANCE." The poem ends with another roll of the dice: "*Toute pensée émit un coup de dés*"—"all thought brings forth a throw of the dice."

"Un Coup de dés" is a powerful statement. At the beginning of the twentieth century Mallarmé was already pushing language to the limit and exploring ways of going beyond the limit by using language in visual ways, by making word and phrase rather than sentence the dominant linguistic element, and by rethinking such fundamental conventions as the relation on the page between figure and ground.

It was, of course, possible to respond to silence with silence. While the prewar avant-garde understood "Un Coup de dés" as a call to creative effort, which is the obvious implication of the echo in its form of the epic journey, it could also be understood as an expression of despair: the bottom has dropped out of things, nothing has meaning any longer. This was the conclusion reached by a great many artists and writers who lived through the First World War. The war seemed to expose the hollowness of traditional values without offering anything to replace them. Silence seemed to be a necessary condition rather than something to be overcome.

Ernest Hemingway's *The Sun Also Rises* nicely illustrates this response. Jake Barnes, the hero of the novel, is a veteran living in Europe after the war. He is impotent—the result of a war wound and also a symbol of what the war has done to his generation. He drifts between France and Spain with a group of rootless expatriates. The old words have been made hollow by the war. The problem is that Jake has not found any new words. He and his friends are adrift without language. They are lost. Stoic silence is close to an admission of defeat, but it is better than speech which always falsifies. The novel is about silence, about the way language cheapens and falsifies, about its inability to express any but the most elementary things and sensations. The most telling comment in the novel is a comment on the value of silence: "You'll lose it if you talk about it."

This is Jake's comment. It sums up his position. He can't talk about his wound. He can't talk about the love he feels for Lady Brett, the novel's heroine. When he does speak, his sentences are laconic. Silence hovers at the edge of everything he says. It eats like a cancer into the lives of the characters. It will destroy them. In the final chapter, Lady Brett makes a hopeful comment: "Oh, Jake . . . we could have had such a damned good time together." Jake's reply is the last sentence in the book: " 'Yes,' I said, 'Isn't it pretty to think so.' "

As a matter of fact, many literary artists not only thought so but did so. In different but related ways, several important twentieth-century literary movements confronted the challenge of silence and managed to have a damned good time in the process. Dada, concrete poetry, and algorithmic poetry all chose this option. If natural languages can be called opaque because they are understood only by a limited group whose members are linked by geography and history, Dada, concrete poetry, and algorithmic poetry share a commitment to discovering modes of language that move toward transparency. Mallarmé's shadow falls on each of them, but each goes beyond him in significant ways.

Since natural languages are geographically localized and historically conditioned in their phonetics, word meanings, grammars, and literary forms, the search for transparent languages seems to deny the nature of language in ways that have no parallels in painting, sculpture, dance, or music. The visual arts seem to adapt almost effortlessly to the idea that expression does not have to conform to historically defined "meaning" in order to be expressive. The same is true of viewers. Today most people who visit art galleries have no trouble appreciating paintings and sculpture that have no meaning in the conventional sense of alluding to historically charged symbols. However, they would probably be shocked if they were asked whether or not language should mean. What is language if not a device for communicating meaning? When language is pushed by various strategies toward transparency, it seems to abandon its capacity to mean in the normal sense of that term. You might say its poetry becomes a poetry of nothing. Is this not a kind of linguistic rape, a violation of the most sacred inner being of the word?

The idea that language does not have to mean is a paradox, but it is not necessarily frivolous. Anyone seriously interested in art today is aware that the languages that emerged in visual art in the early twentieth century were enigmatic only because they were not understood. As people came to understand them, they were seen to be languages with profound lessons to teach. If the point can be readily admitted in the case of visual art, it should not be difficult to admit in the case of literary art.

1 8

THE SEARCH FOR TRANSPARENCY

Science will go wherever logic and experiment take it. Language, on the other hand, is not arbitrary. It is natural, a given, and what is natural is the opposite of what is arbitrary. Infants have no defense against learning a language and they have to learn it the way it is. Only a person who knows one language can refuse to learn another.

Each human mind is formed by a natural language, which, in turn, is a vehicle for the history of a specific culture. The achievement of true originality in thought and expression is therefore extremely difficult. We are always being trapped by the web of associations inherent in language. The point is obvious enough in the case of words. "Father" suggests "mother" and the patriarchal nuclear family; to say "The sun also rises" is to repeat the idea that the sun moves around the earth; "cross" rhymes with "loss"; and even for nonreaders, "to be" is likely to suggest "not to be."

The web of associations is infinitely subtle, and it goes far deeper than noun, verb, and adjective. The linguist Benjamin Lee Whorf points out in "Language and Logic" that the grammar of English and other Indo-European languages requires us to think about reality as a series of subjects distinct from and acting on objects, as in the sentence "John hit the ball." This way of thinking is, he remarks, "natural to Mr. Everyman's daily use . . . of languages," but it is inadequate and even misleading when it tries to cope with "the great frontier problems of science." It hinders as much as it assists speakers of languages like English, French, Russian, Kurdish, and Hindi in their effort to imagine the world revealed by science. If the mind is betrayed by grammar as well as by words, the problem of language is very troublesome indeed.

The problem of language is a by-product of the opacity of natural languages. While English, for example, creates the possibility of imagining one world, it does so at the cost of shutting out an enormous number of alternative worlds, including, if Whorf is right, the world of modern science, which is as close to the real world as modern man has been able to get. There is also the problem of cultural lag. Language is a set of conventions developed laboriously by a social group over generations. Culture, however, has been changing rapidly in the twentieth century. Inevitably, a gap has developed between language and the world; or, to put the idea more pointedly, language increasingly refers to a world that no longer exists. If natural languages separate their speakers from the world because they are opaque, it is natural to ask how language can be made transparent or at least pushed in the direction of transparency.

A step in that direction is to destabilize normal linguistic processes by imposing nonlinguistic rules on them. The most effective rules will be entirely arbitrary—random, like a throw of dice—because any process that is controlled by the artist will bear the imprint of the associations that shape the artist's conscious and unconscious thought and thus veer back toward opacity. Randomness may seem on first consideration to be a desperate strategy, but it is the the central theme of "Un Coup de dés," and it is a recurrent motif in twentieth-century efforts to create new forms in language.

The idea needs to be examined carefully. To be perfectly transparent a language would have to be equally intelligible to all readers or listeners.

No natural languages can begin to meet this condition. Natural languages become increasingly transparent as they are imposed on ever larger geographical areas by imperialism or a melting-pot philosophy, but so far in history, imperialism has created resentments that have eventually halted the spread of every natural language. This is evident in the history of Greek, Arabic, Chinese, Hindi, Spanish, French, English, and Russian.

Ideally, a transparent language should be neutral. It should not be associated with the tyranny of one social group over another and should be equally available to all social groups. Historically, the closest approximation to this limited kind of transparency in the Western world has been medieval Latin. By the time Latin became the universal language of the establishment in medieval Europe it had ceased to be spoken by any of the peoples who used it. It was a dead language, hence an arbitrary—or almost arbitrary—choice. It achieved transparency within its limited geographic region because the culture that had originally produced it had disappeared. Belonging to nobody, it could belong to everybody.

By the same token it had ceased to be a natural language. It had to be taught much as mathematics is taught today. When learned, it became the passport into an international managerial elite whose culture was different in kind from all of the local cultures it administered.

The nineteenth century brought a wave of enthusiasm for what Tennyson in "Locksley Hall" called "the Parliament of man, the Federation of the World." Since there was no possibility of reviving Latin in the Age of Telegraphy, efforts turned to the creation of a synthetic language that might eventually become universal and thus transparent. The most successful product of this impulse was created in 1887 by Ludovic Lazarus Zamenhof, a Polish oculist, who called his language Esperanto—the language of hope. Today there are Esperanto societies throughout the world, and over one hundred journals and newspapers are published in Esperanto. Some six thousand disciples of Zamenhof assembled in Warsaw in 1987 to celebrate the centennial of the movement, and in Europe and North America Esperanto societies are sufficiently active to be able to offer free lessons to anyone who asks for them.

A basic requirement of a universal language is that it be easy to learn, which is one reason why Latin eventually lost out. Zamenhof interpreted this requirement in terms of European languages. The result was a language with a minimum of inflections, a Romance syntax, and a vocabulary as rich as possible in Romance, Germanic, and Slavic cognates. Essential to the language is the regularity of its rules. As in mathematics, there are no exceptions. In this respect Esperanto shows the influence of the scientific point of view.

Esperanto was conceived as a world language for a world culture. As we read in *Esperanto in the Modern World/Esperanto en la Moderna Mondo: "Inteligenta persono lernas la internacian lingvon rapide kaj facile. Esperanto estas la moderna, kultura lingvo por la internacia mondo. La simpla, praktika, fleksebla Esperanto estas la solvo de la problemo de generala interkompreno. Esperanto meritas seriozan konsideran."*

Well, muttered the Devil, it is certainly *simpla, praktika,* and *fleksebla,* but is it transparent?

Even if Esperanto had been more successful than it is, it would have been transparent only in the area of the languages from which its grammar and vocabulary were drawn. There is also a problem utterly unknown to Dr. Zamenhof. It is the one identified by Benjamin Lee Whorf. Indo-European grammar organizes the world in terms of subjects that act on objects. An explanation of the subject-verb relation in a widely used introduction to Esperanto is not so much a lesson in grammar as a lesson in the truth of Whorf's observation: "LANGUAGE IS ALL ABOUT THINGS (nouns) AND THE ACTIONS (verbs) OF ENERGETIC THINGS: ONE THING . . . ACTS ON . . . ANOTHER THING. *Birdo kaptas insektion.*"

Is it true that "*Birdo kaptas insektion*"? Or is that a myth of Indo-European grammar and one of the reasons why Whorf—along with Werner Heisenberg and a great many other modern scientists—despaired of the ability of language to express the realities of particle physics? Is "*Birdo kaptas insektion*" a sentence or a silence or a bald-faced lie? Whorf believed that American Indian languages like Hopi and Navaho offered alternatives to the Indo-European way of putting things together. They are closer, he thought, to the worldview of modern physics because they blur the distinction between subject and object and also the past-present-future distinction so emphatic in the Indo-European verb system. However, American Indian languages are just as opaque as English, and even as they permit one alternative world to be imagined, they preclude others.

Pure nonsense words should be transparent because they are equally unintelligible to everyone unless a word that is nonsense in 4,999 of the world's languages happens to be the name for glaucoma in the five-thousandth. But what looks easy in theory turns out to be very difficult in practice. When creating nonsense words, the mind follows the phonetic conventions of its natural language, so that a German or a Czech or a Japanese or an English nonsense word *sounds* German or Czech or Japanese or English. Even words that are supposed to be imitations of the same natural sound are different in different languages. Dogs say "*bow-wow*" in English, "*ouâ-ouâ*" in French, "*wan-wan*" in Japanese, and "pyee" in Bantu (though only after being kicked). Pigs go "*oink-oink*" in English but "*cué-cué*" in Portuguese, "*snöf-snöf*" in Finnish, and "*fron-fron-fron*" in Italian.

The alternative is to create a language of random sounds. But even sounds conspire to meaning because they tend to follow the patterns of elementary literary forms—lament, celebration, gratification, rejection, and the like—with the further qualification that even elementary literary forms tend to be specific to individual language groups. On the other hand, if the sounds can be combined as well as selected by random processes, they will have no culturally determined form and hence should be perfectly transparent.

At least when uttered. When they are printed, a whole new family of problems arises because some alphabets are phonetic and others are pictographic. Printing sounds in a Roman alphabet localizes them just as much as printing word-pictures in Chinese characters. A perfectly transparent printed poem would be in an alphabet as randomly created—hence as unintelligible—as the sounds themselves.

This line of thought leads to two conclusions: first, that natural languages are irrevocably tied to history, and second, that the idea of scrapping history is an illusion. The only way to escape history is to compose unintelligible, randomly selected sounds in an alphabet that has never been used, which is absurd.

The story of efforts to create transparent languages is therefore a story of compromise. Consider nonsense verse. It is intriguing and probably significant that the first self-conscious efforts to create nonsense occurred during the rise of industrialism and that the most successful English poet in this vein was a logician. Lewis Carroll's solution to the problem was to introduce nonsense words into otherwise conventional English sentences. His poems are delightful. However, they are imperfect as nonsense. Even though the words have no dictionary meanings, they become, in context, recognizable subjects and objects and verbs and predicates and the like for the reader who is familiar with English. Take Carroll's poem "Jabberwocky":

> And as in uffish thought he stood,
> The Jabberwock with eyes of flame,
> Came whiffling through the tulgey wood,
> And burbled as it came.

In spite of the nonsense words, any English reader knows immediately that grammar is hard at work in the lines quoted. Although the reader cannot say exactly what a Jabberwock is, it is clearly a noun singular, subject of *came*. There is more. The grammar of "Jabberwocky" is recognizable Indo-European. The subject is out there all by itself, quite absolutely separated from all potential objects and also from the tulgey wood in which it is whiffling. There is still more. Because the poem in which this stanza occurs is based on a conventional literary form—the knightly quest—the reader concludes that the Jabberwock is a monster that has to be slain by the hero. He knows, too, that the Jabberwock inhabits a tulgey wood. It has the characteristics of an animal, and its whiffles and burblings are both sinister and delightful. They reinforce the convention and the parody.

"Jabberwocky" is amusing, but the humor will be mostly lost on readers who do not speak English and even more on speakers of languages whose grammars carve up the world in ways different from the grammars of Indo-European languages. It will be lost most of all on speakers of languages whose repository of genres does not include something like the knightly quest. Apparently, a perfectly transparent language must be emptied of all associations with natural and local languages.

Is this possible in language? If any literary movement of the twentieth century furnishes clues to answer this question, it is the movement with the mysterious and resonant name "Dada."

19

DADA

Dada began in Zurich in 1916 when Hugo Ball, poet and philosopher, and Emmy Hennings, singer, began a series of "cabarets," eventually called the "Cabaret Voltaire," at a bar called Hollandische Malerei. A call for performers attracted Tristan Tzara along with Hans Arp, later famous as a sculptor, and Marcel Janco, poet and painter. The first cabaret was held on February 5. Posters announcing the performance associated the new movement with futurism, but Ball, Tzara, and company soon parted company with futurism in favor of their own more radical program. Richard Huelsenbeck, a poet, became part of the cabaret in February, and the central group was assembled. Its progressively wilder performances included simultaneous recitations of banal texts in several languages, poems composed of sounds without meanings, raucous banter between performers and audience, and much song and dance.

Such was the success of Ball's venture that the Cabaret Voltaire was shut down by public demand on June 23, 1916. By then the movement had acquired its name—Dada—thanks (according to Ball) to a random stab of a knife blade into a Larousse dictionary. It would continue to flourish, most luxuriantly in Zurich, Paris, and Berlin, until 1923. Among the poets and artists, other than the founders, associated with Dada were Kurt Schwitters, Francis Picabia, Philippe Soupault, Jean Cocteau, Marcel Duchamp, Man Ray, and André Breton. The end came at a performance of Tzara's play *Le Coeur à Gas* in Paris in July, 1923. A battle-royal erupted between followers of Tzara and followers of Breton. Police were called in, and Dada was over as a movement.

There are any number of interpretations of the word *Dada*: aesthetic, philosophical, linguistic, sociological, political, psychological, and more. In *Dada: Monograph of a Movement* (1957), Willy Verkauf argues that it was a protest against the "senseless mass murder" of the First World War, and "the hectic outcry of the tormented creature in the artist, of his prophetic, admonishing, tormented conscience." This is impressive rhetoric, and it is clear that the background of a senseless war influenced all of the founders of Dada, including Tristan Tzara. However, Verkauf's emphasis on the "tormented artist" seems rather at odds with the account that Tzara gives of the playful atmosphere in which Dada was born. The same point may be made of Hans Knutter's suggestion that Dada was a form of infantile regression from the horrors of the adult world, and the commonplace assertion that Dada was motivated by a perverse desire to upset the middle class (*épater les bourgeois*).

Many of the Dada artists, including Tzara, became political activists, and several of them have argued that Dada was political from the beginning. Arp, conversely, argues in *On My Way* (1948) that the basic thrust was always the liberation of the individual. His comment is obviously touched by a political animus as strong as that of the leftists, but it deserves to be quoted: "Some old friends from the days of the Dada campaign, who always fought for dreams and freedom, are now disgustingly preoccupied with class aims. . . . Continuously they mix poetry and the Five Year Plan in one pot; but this attempt to lie down while standing up will not succeed. . . . It is hard to explain how the greatest individualists can come out for a termite state."

Tzara himself claims the basic attraction of the word *Dada* was that it has no meaning: "DADA MEANS NOTHING," he wrote in his *Manifesto* of 1918. It means nothing, and nothing is the true subject of Dada. In the same *Manifesto* Tzara attacks the ideologists who try to find meanings where no meanings exist. The strategy used to find meaning is to attach a history to Dada: "The first thought that occurs to these people is bacteriological in character: to find [Dada's] etymological, or at least its historical or psychological origin. We see by the papers that the Kru Negroes call the tail of a holy cow Dada. The cube and the mother in a certain district of Italy are called: Dada. A hobby horse, a nurse both in Russian and Rumanian: Dada. Some learned journalists regard it as an art for babies. . . ." All efforts to attach Dada to history contradict its intention. It has no history. It is, rather, an art of images without content and words without meaning.

Nothing is not used by Tzara to mean nihilism in the sense of a bitter and often despairing rejection of traditional values, but to identify the liberating power of the arbitrary. Dada shares its interest in the arbitrary with modern science, and both frequently express the sense of the freedom associated with the arbitrary through playfulness and the metaphor of creation as game. In becoming nothing, poetry escapes the prison of convention—of ideology, of flaccid conventions of diction, of outworn canons of the beautiful, and the like—and can become everything simply because it is not tied to one specific set of meanings. Tzara writes: "Freedom: Dada Dada Dada, a roaring of tense colors, and interlacing of opposites and of all contradictions, grotesques, inconsistencies: LIFE."

This brings us directly to the poetry of nothing. A Rorschach ink blot is not nihilistic. It is nothing. Its lack of meaning is its value. The viewer fills it with the contents of his or her subconscious, just as the Eiffel Tower manages to be all things to all viewers by virtue of its freedom from historically defined architectural form. No one objects to white plaster. It is only mosaics of Christ Triumphant and paintings of workers rebelling against capitalist masters that have to be covered up—by Moslems in Constantinople and captains of industry in Rockefeller Plaza in New York. Nor does anyone consider white walls nihilistic. They seem cool, abstract, and inviting, like shimmering silver high-rise buildings of the late International Style in Los Angeles.

The unique value of a laissez-faire economy is that it restricts nobody.

You might call it an economy of nothing, controlled, in Adam Smith's metaphor, by an invisible hand. By allowing free rein to the avarice of everybody, it ensures the best of possible economic worlds. Nor is the invisible hand nihilistic. It is only invisible. Its invisibility is its most significant feature. It is so invisible that nobody had seen it before Adam Smith gave it a name.

In architecture the equivalent of the nothingness of Dada is the open floor plan. Buildings with open floor plans have hung ceilings and Sheetrock partitions. Everything is temporary. The interiors *look* solid to the visitor, but they have no serious commitment to the past. Throne rooms have to be blown up after a revolution, but a week after the bankruptcy of a great corporation—or, alternately, a sudden business expansion or a leveraged buyout—the company boardroom has been swiftly and painlessly converted into an office for a parachute consultant. Seen from the perspective of Dada, the open floor plan succeeds because it has liberated itself from history. It is a nothingness, a kind of transparency.

Tzara's 1918 manifesto anticipates all of its possible challenges: "I write a manifesto and I want nothing . . . in principle I am against manifestos, as I am also against principles." Philosophy is turned out in "factories of thought." Logic is a disease of the mind. Experience is a product of chance, and the message of science is make love and bash your brains in. All system is systematic falsification. Art always turns into posturing. Although the year is only 1918, Tzara announces that the revolution begun by Picasso and Braque has already been coopted: "We have enough Cubist and futurist academies." Meaning is a cop-out. Recognition of a writer by the press is "proof of the intelligibility of his work: wretched lining of a coat for public use."

We are concerned here, however, with language rather than fashions in art. Dada seeks transparency through combinations of words from natural languages with nonverbal modes of expression. The problem is that it appears incapable of making statements about the human condition that can be considered significant. "Appears" is the right word because Dada may be expressing something we still understand only imperfectly—namely, the separation of the mind in technological society from history. If so, one significant idea that Dada expresses is the movement of humanity beyond nature and toward the habitat of scientific universals. In other words, it enacts a process of disappearance—of a making-of-the-transparent.

Cover of 391 *(#12, 1920). When pronounced, the letters* L H O O Q *are intended to make a sentence: "Elle a chaud au cul."*

A key Dada strategy for creating transparency is randomness. Much Dada poetry is made by one or another equivalent of rolling dice. Achieving randomness, however, is more difficult than it sounds and than the initial Dada experiments suggest. In certain areas of high technology randomness is essential. Consequently the subject has been explored in depth. All simple approaches to the problem result in patterns. A typical method of constructing a Dada poem, for example, is cutting up the phrases or words in a newspaper editorial, stirring them in a hat, and pasting them on a piece of paper in the order in which they are withdrawn. Obviously, the words come in this case from a natural language, and it is equally obvious that, being printed, they are in a visual medium (the Roman alphabet) that is opaque rather than transparent. Even if you ignore these objections, you still confront the fact that the words are an inadequate statistical sample of the words in the natural language from which they were drawn.

Far from being an arbitrary gathering, they reflect the idiosyncrasies of the author, the expectations of the audience to which they were addressed, the subject matter of the article, the state of the language when it was written, and the like. All of these factors are determined by history. An article on the culture of grapes from a French newspaper of the 1880s, for example, uses a vocabulary quite different from an article on the same subject from a technical journal published in 1985. Words clipped from *Time* magazine's article on the death of Elvis Presley will have a different bias from words clipped from an article on Benjamin Franklin's Parisian dalliance in *Proceedings of the American Antiquarian Society* for the same year.

All strategies for achieving randomness in language have defects. Decisions about word choice might be made by flipping a coin or rolling dice in the manner of Mallarmé's poem or by using the *I ching* in the manner of John Cage. But coins and dice are biased and become more so as they are used, and one of the special values attributed by Cage to the *I ching* is its ability to reveal significant patterns—to be the opposite of random. Randomness is an unattainable ideal. In fact, there is a question of how one could recognize perfect randomness if one confronted it.

These facts are remote from everyday experience and not immediately relevant to Dada, but they define the limits of the Dada effort to achieve transparency.

Hugo Ball claims to have invented one of the purest and most translucent forms of Dada, the sound poem. He explains it as follows:

> I invented a new species of verse, "verse without words," or sound poems. . . . I had a special costume designed for it. My legs were covered with a cothurnus of luminous blue cardboard, which reached up to my hips so that I looked like an obelisk. . . . I recited the following:

> gadji beri bimba
> glandridi lauli lonni cadori
> gadjama bim beri glassala
> glandridi glassala tuffm i zimbrabim.
> blassa galassasa tuffm i zimbrabim. . . .

> . . . With these sound poems we should renounce the language devastated and made impossible by journalism. We should withdraw into the innermost alchemy of the word, and even surrender the word, in this way conserving for poetry its sacred domain. . . . We should no longer take over words . . . which we did not invent absolutely anew, for our own use.

"Gadji beri bimba" illustrates the limits of transparency when all natural words have been abandoned. Its phonetic values and regularities point to a European imitation of an unknown African language. The strong use of alliteration and consonance and the emphatic rhythms further identify the European tradition within which the poem was composed. One might speculate that *galassasa* is a noun, *tuffm* a verb, *i* an article, and *zimbrabim* another noun, object of *tuffm.* The poem thus traps its reader in Benjamin Lee Whorf's Indo-European shooting gallery where subjects are always zeroing in on objects. These defects aside, the poem is a persuasive experiment in creating language without meaning.

Kurt Schwitters composed *Ursonata,* a sound sonata in four movements. A more concise example of his art is "W," a poem of one letter. In *Vision in Motion* (1947), Laszlo Moholy-Nagy describes Schwitters's recitation of the poem: "He showed to the audience a poem containing only one letter on a sheet: W. Then he started to 'recite' it with slowly rising voice. The consonant varied from a whisper to the sound of a wailing siren till at the end he barked with a shockingly loud tone."

Moholy-Nagy uses primitive words to describe this recitation: *whisper, wailing, barked.* He is not alone in his observation that at its most experimental, avant-garde poetry is also primitive. Ball recalls that during the recitation of his own sound poetry, "I . . . noticed that my voice, which seemed to have no other choice, had assumed the age-old cadence of the sacerdotal lamentation, like the chanting of the Mass."

Ball's "gadji beri bimba," in other words, resembles a common religious

SONATE

Grim glim gnim bimbim
grim glim gnim bimbim
grim glim gnim bimbim
grim glim gnim bimbim
grim glim gnim bimbim
grim glim gnim bimbim
grim glim gnim bimbim
grim glim gnim bimbim
bum bimbim bam bimbim
bum bimbim bam bimbim
bum bimbim bam bimbim
bum bimbim bam bimbim
grim glim gnim bimbim
grim glim gnim bimbim
grim glim gnim bimbim

grim glim gnim bimbim
bum bimbim bam bimbim
bum bimbim bam bimbim
bum bimbim bam bimbim
bum bimbim bam bimbim
Tila lola lula lola
tila lula lola lula
tila lola lula lola
tila lula lola lula
Grim glim gnim bimbim
grim glim gnim bimbim
grim glim gnim bimbim
grim glim gnim bimbim
bem bem
bem bem
bem bem
bem bem

Tata tata tui E tui E
tata tata tui E tui E
tata tata tui E tui E
tata tata tui E tui E
Tillalala tillalala
tillalala tillalala.
Tata tata tui E tui E
tata tata tui E tui E.
Tillalala tillalala
tillalala tillalala.
Tui tui tui tui tui tui tui tui
te te te te te te te te
tui tui tui tui tui tui tui tui
te te te te te te te te.
Tata tata tui E tui E
tata tata tui E tui E.
Tillalala Tilla lala
tillalala tilla lala

Tui tui tui tui tui tui tui tui
te te te te te te te te
tui tui tui tui tui tui tui tui
te te te te te te te te
O be o be o be o be
o be o be o be o be.

KURT SCHWITTERS

Kurt Schwitters, Sonate *(1919)—a short sound poem in the same mode as* Ursonata.

phenomenon—glossolalia, or speaking in tongues. The "W" poem of Kurt Schwitters moves in a similar way toward an approximation of what is evidently a primal scream. This is analogous to the tendency of simple geometric objects—a tower, a sphere, a circle—to be interpreted by viewers in terms of sexual or mythic fantasies. As it sheds meaning, language begins to touch the universals of speech, and this universality is an aspect of its transparency.

The point, however, is secondary. The basic object of sound poetry is identified by Moholy-Nagy in another comment on Schwitters's "W" poem:

> The only possible solution [to the stifling of poetry by convention] seemed to be a return to the elements of poetry, to noise and articulated sound, which are fundamental to all languages. Schwitters realized the prophecy of Rimbaud, inventing words "accessible to all five senses." His poem *Ursonata* (1924) is a poem thirty-five minutes in duration. . . . The words do not exist, rather they might exist in any language; they have no logical, only an emotional context.

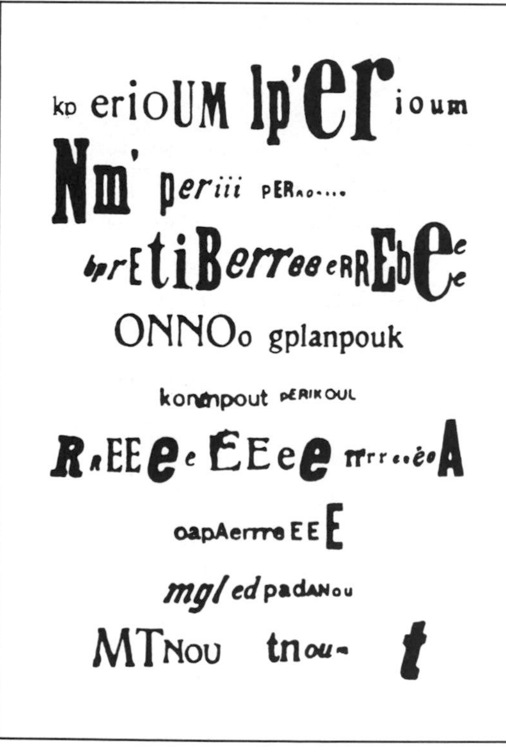

Raoul Hausmann, "Optophonetic Poem" (1918).

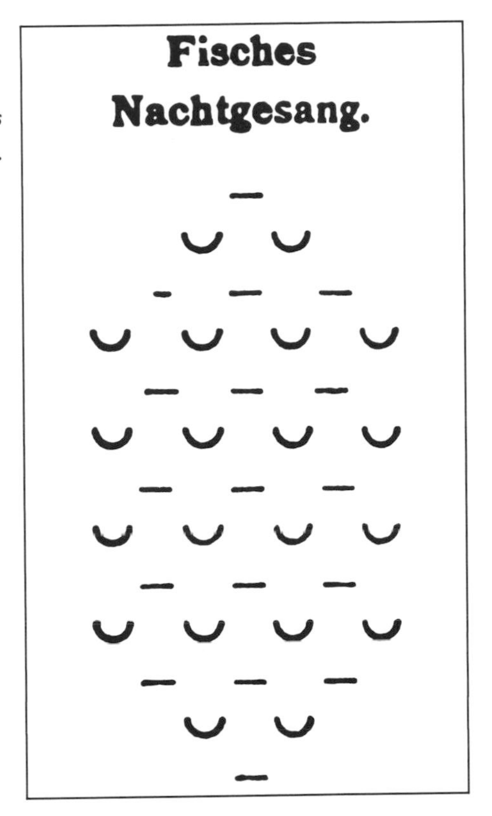

Christian Morgenstern, "Fisches Nachtgesang," phonetic poem (1920).

A phonetic poem in which image absorbs sound is another effort to achieve linguistic transparency. Raoul Hausmann composed "Optophonetic Poem" in 1918; it consists of letters in various type faces and creating pronounceable syllables (see p. 174). The first line is "kp'erioum Ip'erioum." This is not prepossessing but becomes so because of the striking visual presentation. Hausmann has complained that *Ursonata* borrowed some of its ideas from "Optophonetic Poem," but that is another argument. Christian Morgenstern's "Fisches Nachtgesang" ("Fish's Nightsong") dispenses with syllables and letters entirely (see above). On the other hand, the series of flat and U-shaped marks that constitute it are scansion symbols used to mark poetic meters, so it remains culture-bound. In 1924 Man Ray composed a phonetic poem of short and long lines in a notation that looks like Morse code but is perfectly without meaning. He bowed to history, however, by making the long and short lines regular in length and arranging them in the shape of a title followed by four stanzas.

A famous Dada technique for achieving transparency in language was clipping words from journals. The effect is clearly more striking for noninflected languages like English than for inflected languages like Latin, where the inflections allow word order to be relatively permissive. Tzara gives the formula for this kind of poetry and a sample poem:

> To make a Dadist poem
> Take a newspaper.
> Take a pair of scissors.
> Choose an article as long as you are planning to make your poem.
> Cut out the article.
> Then cut out each of the words that make up the article and put them in a bag.
> Shake it gently.
> Then take out the scraps one after the other in the order in which they left the bag.
> Copy conscientiously.
> The poem will be like you.
> And here you are a writer, infinitely original, endowed with a sensibility that is charming, though beyond the understanding of the vulgar.
>
> Example:
> When the dogs cross the air in a diamond like the ideas and the appendix of the meanings show the hour of the awakening program. [This title is my own.]
>
> price they are yesterday agreeing afterwards paintings Appreciate the dream epoch of the eyes pompously than recite the gospel mode darkens group the apotheosis imagine he said fatality power of colors . . .

Words, however, are less transparent than sounds or straight lines. In spite of the lack of grammar and syntax, Tzara's words trap the reader in their natural meanings. The literary critic buried in every psyche longs to discover their meaning through clever exegesis or expose them as barefaced fraud. In either case, the lesson is lost. The threat of the new posed by the disorientation is contained. In the first instance, the new is subtly changed into the familiar; in the second the problem is resolved by dismissal.

In *Science and the Modern World*, Alfred North Whitehead observes that it takes an extraordinary intelligence to contemplate the obvious. One of the prime tasks of poetry—it may be *the* prime task—is to contemplate the obvious. In the famous definition of poetry that Samuel Taylor Coleridge offers in Chapter 14 of the *Biographia Literaria*, the basic appeal of poetry is said to be its capacity for "awakening the mind's attention from the lethargy of custom and directing it to the loveliness and the wonders of the world about us; an inexhaustible treasure, but for which, in consequence of the film of familiar and selfish solicitude, we have eyes, yet see not, ears that hear not, and hearts that neither feel nor understand."

Whether or not the poets of Dada were revealing unappreciated beauties and wonders, their fascination with typography and newspapers suggests that they were contemplating the obvious transformation which language was undergoing in the early twentieth century in the mass media. Marcel Duchamp had presented commonplace, commercially produced objects—for example, a bicycle wheel and (most notoriously) a urinal signed "R. Mutt 1917"—as though they were works of art. He called them "ready-mades." Today they would be called "found objects." Found objects are entirely respectable. They are continuous with a perhaps more interesting phenomenon—the entry into museums of such products of modern culture as typewriters, space capsules, and Stutz Bearcats.

Is it not a literary manifestation of the same impulse to make poems out of materials clipped from newspapers in the manner of André Breton? In fact, does not a modern newspaper have Dada characteristics? To test this theory one need only try to imagine how today's newspaper would appear to a reader unfamiliar with the conventions of modern journalism. It would seem on first glance to be a haphazard mosaic of columns of different lengths, of photographs, of headlines in varying sizes of type and column width, and of advertisements. Trying to "read" this mosaic, the naive reader would find himself following a story about political corruption in Manila, then breaking off to read a report about *glasnost* in Murmansk, then breaking off to a story on a teenager who died in a fire, breaking again to read about election fraud, then breaking off again . . . and so forth. The only element connecting the stories on the front page is coincidence: they all happened within twenty-four hours before the paper was printed.

The front page is thus unpredictable, a product of randomness just as much as a Dada poem clipped from a newspaper. Moreover, the randomness is discontinuous. Seldom do stories end on page one. Simply because they are important enough to make it to page one, they are usually continued by jumps to later pages. But most people read the front page all at once. Thus they gather fragmentary impressions of several disconnected events rather than a full impression of a single event, as one does when reading a book or (for the most part) a magazine article.

Discontinuity and fragmentation are part of the deep structure of modern culture. They are visually expressed by collage. The equivalent use of language is nicely illustrated by a passage from Marinetti's *Variety Theatre* (1913): ". . . nostalgic shadows besiege the city brilliant revival of streets that channel a smoky swarm of workers by day two horses (30 metres tall) rolling golden balls with their hoofs GIOCANDA PURGATIVE WATERS crisscross of *trrr trrrrr* Elevated *trrrr trrrrr* overhead trombone whistle ambulance sirens. . . ." Similar collage effects are used in *Ulysses* by James Joyce to reproduce the flow of ideas through the mind of Leopold Bloom. Joyce does not intend a vision of the future. He is trying to represent the condition of life in a grubby and entirely commonplace corner of the everyday world.

Today the discontinuities objectified in newspapers and imagined by Joyce have been dwarfed by television. An hour of commercial television programming may begin with the title of a drama and open with five minutes of a touching scene in which a daughter speaks with a mother who is dying of cancer. With minimal warning, the program cuts to an advertisement for a hemorrhoid remedy: Preparation H. Then to a newsbrief: "Light snow expected in the metropolitan area early this evening; traffic advisory on the Beltway." Next a girl in a bathing suit spirals down through space until she lands in the driver's seat of a Honda Accord, while a disembodied voice announces that 6.8 percent financing is available until Monday. Then cut to the cancer ward and the daughter weeping over the dying mother.

When Jean Cocteau used abrupt discontinuities in his surrealist film *Orpheus* the art world was enchanted. How advanced, how outrageous! The discontinuities of *Orpheus* are trivial compared to the discontinuities accepted as the normal mode of television by T.V. afficionados of the developed world. The psychoanalytic surrealism of *The Cabinet of Dr.*

Caligari or of Ingmar Bergman's *Wild Strawberries* is timid compared to the surrealism that the teenagers ingest as a daily diet from musical videos, to say nothing of the spectacular happenings that have become standard fare at concerts by popular entertainers like Michael Jackson or Kiss or Madonna.

The pattern of continuous discontinuity can be traced across the whole spectrum of modern culture. It is a reflex of two of the driving forces of that culture: first, of technology, which exists to innovate, so that the breakthrough of one year becomes the anachronism of the next; and second, of the dependence of the industrial economy on the rapid obsolescence of everything it produces, from stockings to automobiles to movies. In modern culture, all art is temporary art and happenings replace ceremonies. That's how jobs are created. Again, the newspaper can be seen to be an imitation of the conditions of its culture: it is instantly obsolete. Indeed, nothing is more troublesome in a modern household than the mass of newspapers that accumulate during the week while they await disposal. Elderly people are sometimes found in houses so crammed with newspapers waiting for disposal that no room is left for them to move in. Because of obsolescence, newspaper headlines can become pure Dada a few weeks after they have appeared. Consider the following, randomly chosen from an American newspaper published in the late 1970s:

Suburbs Push Spartan Water Habits
Thurmond's Switch
Old Celebrations, New Translations, Gossip, and Ghosts
Nambi's Dunes Hide Wealth of Diamonds
Energy in August
Orioles Toppled by Rare Blasts from Nordhagen
Sounders Earn Date With Cosmos
Weaver Gives Thumb Again

Tzara and his friends were not inventing something new; they were merely contemplating the obvious and imitating what had already arrived. Imitation has been recognized as the task of art since Aristotle's *Poetics*, so the idea can hardly be called radical.

20

CONCRETE POETRY

By 1939 Dada had disappeared from the European literary firmament, a victim of the general sense that life was real and life was earnest and about to get a lot more so. The reaction began in the Soviet Union with repudiation of the early ideals of the revolution and subsequent liquidation by Stalin of the intelligentsia, including avant-garde poets and constructivist artists and architects. It spread to the Western democracies in the 1930s along with the Depression, and the shadows cast by the approach of the Second World War turned retrenchment into headlong retreat.

During the period of retrenchment, art was summoned to the barricades by ideologues of all persuasions. Many avant-garde artists turned to the left along with André Breton, Louis Aragon, and Tristan Tzara; others turned to the right along with Marinetti. In the midst of all the political activity, and even as the leaders of public taste denounced it, modern art continued the process of finding ways to overcome the silence that numbed such a large part of the world. Many artists and much art found refuge in America, and much of the modern tradition was passed by them to a new generation of American artists. The Bauhaus was particularly successful in accommodating itself to the New World, and, perhaps surprisingly, American popular culture welcomed the modern aesthetic as presented in advertisements, graphics, poster art, and design. Whatever its detractors might claim, the new art communicated.

The end of the Second World War was a liberation in more than one sense in Western Europe and the United States. The censors and the secret police were, for the moment, thoroughly discredited, and for a few years there seemed to many to be a real possibility that a new world order might be established. Concrete poetry—*poésie concrète*—was born out of that moment of optimism and incorporated into its program both the internationalism of the avant-garde movements of the 1920s and their interest in new languages able to express the realities of life in an industrial and technological society.

The emphasis of concrete poetry is on language as image. It seeks to make language expressive by giving it a visual element rather than by forcing it into new shapes by dislocations of normal vocabulary and grammar in the manner of Dada poetry. The idea that a text should be visually expressive recalls Mallarmé, but it also recalls much earlier periods of communication by writing. It may seem to be anticipated by the tradition of illuminated manuscripts, but it is not, and the difference is instructive. An illuminated manuscript is usually a text that has been illustrated. At best it is the result of a close collaboration between artist and writer. Even in the best examples, however, word and image remain separate.

Concrete poetry advocates a merging of the two domains. In this it is seeking to recover a unity lost in Western culture but still present in Chinese, where many of the characters used for writing retain an explicit relation to objects in the world. Hieroglyphics have a similar visual aspect, and memories of a primitive relation between the alphabet and the world are preserved, for example, in the Hebrew alphabet, where the first letter—"aleph"—means "ox," the second—"beth"—means "house," and the third—"gimel"—is a still-audible "camel." When writing con-sisted of shapes and sounds related directly to common objects, it had a certain inherent transparency. People could understand how it worked even if they could not read what it said. When letters became pure phonetic symbols, however, the transparency was lost. Concrete poetry seeks to restore it by making words images of what they are saying. In the process, it also creates the possibility for the word to *be* something without having to mean something.

Visual and shaped poetry delighted the ancient Greeks and has been written ever since. A shaped poem is written so that its lines form an image. Normally the shape is a comment on the subject. A religious poem takes the shape of an altar or a cross. A poem about fire has the shape

"Fury said to
a mouse, That
he met
in the
house,
'Let us
both go
to law:
I will
prosecute
you.—
Come, I'll
take no
denial;
We must
have a
trial:
For
really
this
morning
I've
nothing
to do.'
Said the
mouse to
the cur,
'Such a
trial,
dear sir,
With no
jury or
judge,
would be
wasting
our breath.'
'I'll be
judge,
I'll be
jury,'
Said
cunning
old Fury;
'I'll try
the whole
cause,
and
condemn
you
to
death.'"

"Mouse Tale," from Alice's Adventures in Wonderland *(1865).*

of a flame. Poems take the shape of axes, pillars, pyramids, diamonds, eggs, and spears. Among the most admired poems of George Herbert (d. 1633) is one whose shape is described by its title: "Easter Wings." Surely, however, the best anticipation of modern concrete poetry, both in its zany use of typography and its wit, is Lewis Carroll's representation of the way that Alice heard the Mouse's tale while looking at its tail (facing page).

The first explicitly concrete poems were probably the *Testi-poemi murali* of Carlo Belloli, begun in 1944. The movement, however, is usually traced to Ernst Gomringer, a Bolivian-born poet working in Switzerland. Around 1950 Gomringer found that his writing had come to a dead end. Having been asked to be literary editor of a new magazine, *Spirale*, devoted to poetry and the visual arts, he began experimenting with visual and typographic effects and in 1952 produced his first concrete poem "Avenidas," which consists of three words:

avenidas
avenidas y flores

flores
flores y mujeres

avenidas
avenidas y mujeres

avenidas y flores y mujeres y
un admirador

Ernst Gomringer, "Avenidas" (1952).

Shortly thereafter Gomringer composed a more austere poem—"Silencio"—consisting of only one (or perhaps two) words:

silencio silencio silencio
silencio silencio silencio
silencio   silencio
silencio silencio silencio
silencio silencio silencio

Ernst Gomringer, "Silencio" (1953).

It is a metaphysical poem. The missing word—the silence announced by the absence of "silence"—is the word that defines the poem and the theme that it expresses, which is also the central theme of poetry since Mallarmé. Gomringer calls his collection of poems "Constellations." The title is an allusion to the constellation that appears just before the end of "Un Coup de dés."

At more or less the same time that Gomringer discovered concrete poetry in Switzerland, a group of Brazilian writers including Haroldo de Campos and Décio Pignatari began using similar strategies. After the Brazilian and Swiss writers learned of each other's work, they discovered similar experiments elsewhere. Evidently concrete poetry was not an isolated phenomenon but a general trend toward a new kind of world literature objectifying an international consciousness. Poets in the United States, England, Austria, Sweden, Turkey, Japan, and many other countries were experiencing the same problems and coming up with similar answers.

This simultaneity is extraordinarily interesting. It parallels earlier twentieth-century explosions of awareness that almost immediately became international—in science with relativity and quantum mechanics, in the visual arts with futurism and Cubism, in architecture with the Bauhaus, in literature with Dada and modernism. Such recurrent explosions of consciousness do not seem accidental. They are like a volcanic eruption that suddenly releases enormous energy after a long, invisible buildup of pressure. They suggest the presence of a form of consciousness just under the surface of modern life that emerges whenever the conditions are right.

Concrete poetry does not share the interest of Dada in nothing and is generally indifferent to the sensational "happenings" that Tzara and Duchamp so much enjoyed. On the other hand, many concrete poets agree with the Dada argument that language in its traditional forms is dead and thus incapable of expressing living ideas and feelings. In 1920 Piet Mondrian, Theo van Doesburg, and Anthony Kok published what might well stand as a manifesto of this view of language. A translation by Mike Weaver appeared in *The Lugano Review* of 1966:

THE WORD IS DEAD . . .

THE WORD IS IMPOTENT
asthmatic and sentimental poetry
the "me" and "it"
which is still in common use
everywhere . . .
is influenced by an individualism fearful of space
the dregs of an exhausted era . . .

psychological analysis
and clumsy rhetoric
have KILLED THE MEANING OF THE WORD

To restore language to its once-proud status, the authors of the manifesto offer a program that anticipates many aspects of concrete poetry: reinvention of syntax, spelling, rhythm, and typography, and a new emphasis on the word as pure sound.

After two decades of ideology and in the midst of the dreary repetition of slogans of the right and left which came with the Cold War, language threatened to become as impotent in the 1950s as it was when Mondrian, van Doesburg, and Kok wrote their manifesto.

One alternative to the asthmatic impotence of words is a language as pure as mathematics. The Czech concrete poet Zdenek Barborka argues in his essay "New Poetry" (1967; tr. Hana Benes) that there are two human natures—a "natural" human nature reflected in history, and a rational human nature reflected in science. Natural human nature is corrupt and debased. The way out of the quagmire is "the unnatural [i.e., scientific] road, which has given [man] the human qualities that have enabled him to create the greatest works of civilization, and which is able to replace human nature entirely." Science is not dehumanizing but enabling: "The moving force of a strictly rational and speculative method is, in the last analysis, a deep love of the world."

Barborka's concept of a poetry based on the values of science repeats a distinction common in the literature of concrete poetry between Expressionistic and Constructivist styles. Expressionistic poems are rooted in natural language and its meanings and are thus trapped in the pastness of language. Barborka would consider them creations of natural (and thus corrupt) human nature. Concrete poems in the Constructivist mode avoid the trap by rising to the level of abstraction that characterizes science. In 1958 *Noigandres,* the Brazilian journal of concrete poetry, explained that Constructivist poetry seeks "pure structural movement . . . ; at this phase, geometric form and mathematics of composition (sensible rationalism) prevail."

The relation between concrete poetry and science complements the world aspirations of concrete poetry. Marinetti wrote in the introduction to Belloli's *Testi-poemi murali:* "These text-poems anticipate a language of word-signs set in the communications network of a mathematical civilization." The epigraph of Ernst Gomringer's journal *Konkrete Poesie/ Poesia Concreta* is: "Concrete poetry is the aesthetic chapter of the universal linguistic form of our epoch." Gomringer argued that concrete poetry expresses "the contemporary scientific-technical view of the world. . . . I am therefore convinced that concrete poetry is the process of realizing the idea of a universal poetry." Mary Ellen Solt, one of the most knowledgeable American authorities on the subject, agrees. The title of her 1970 book is explicitly transnational: *Concrete Poetry: A World View.*

The desire to create a form of expression as pure as science and as universal as a transparent language explains those concrete poems that seek to become pure visual experience, emptied of dictionary meaning. Poems, Belloli argued in *Testi-poemi murali,* should be "anonymous, silent, almost invisible: words set free in an almost transparent medium." There is no reason for the poet to be limited to words, and in fact the poet is most poetic when inventing languages. Hence the concept of the poet as "language designer." Décio Pignatari argued in 1964 that language is "any set of signs and the way of using them." Diagrams, traffic signs, map symbols, mathematics are all designed languages, and interest in designed languages relates concrete poetry to the growing influence in both Europe and the United States of semiotics—the theory of signs.

Mary Ellen Solt's "Moonshot Sonnet" is an example of a poem that creates a language of visual form. Its content, she explains, is formed from symbols used by NASA to mark photographs of the moon. The symbols create the familiar 4–4–3–3 structure of the sonnet. The poem seems to be related in concept to Morgenstern's "Fisches Nachtgesang" and thus to Dada. However, Solt relates "Moonshot" to the aims of the concrete poetry movement. She notes that the sonnet is "supra-national, supra-lingual," and she also stresses that it is "about" a very current development in technological culture—space travel:

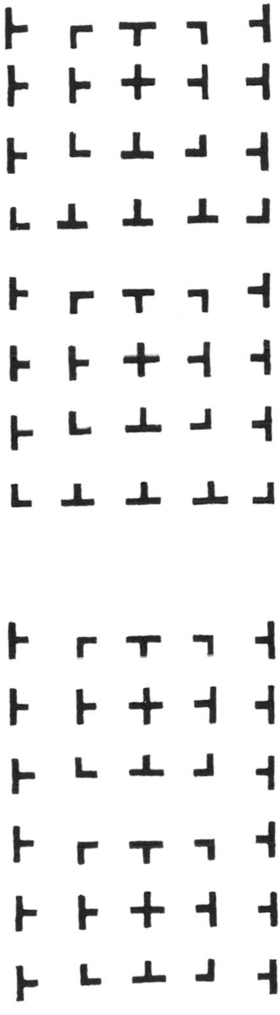

Mary Ellen Solt, "Moonshot Sonnet" (1964).

├ ┌ ┬ ┐ ┤
├ ├ ┼ ┤ ┤
├ └ ┴ ┘ ┤
└ ┴ ┴ ┴ ┘

├ ┌ ┬ ┐ ┤
├ ├ ┼ ┤ ┤
├ └ ┴ ┘ ┤
└ ┴ ┴ ┴ ┘

├ ┌ ┬ ┐ ┤
├ ├ ┼ ┤ ┤
├ └ ┴ ┘ ┤
├ ┌ ┬ ┐ ┤
├ ├ ┼ ┤ ┤
├ └ ┴ ┘ ┤

Alternately, a poem can aspire to the condition of music by evoking sound in the manner of Matthias Goeritz's "Echo of GOLD," in which sound becomes visual. The "poem" is a wall made of large metallic letters, gold in color, spelling "GOLD" over and over in the manner of an echo. Ladislav Novák's "Magic of a Summer Night" also aspires to music. The text, while retaining a strong visual element, is really a musical score for voice:

tma	tma	tma	tam
tma	tma	tma	tma
tma	tam	tma	tma
tma	tma	tma	tma
tma	hma	tma	tma
tma	tma	tám	tma
tam	tma	tma	tma
tma	tma	tma	tam
srp	srp	srp	srp
srp	srp	srp	prs
srp	prs	srp	srp
srp	srp	srp	srp
prs	srp	srp	srp
srp	srp	pes	srp

Ladislav Novák, "Magic of a Summer Night."

Much concrete poetry is Expressionist. Like modernist poetry, it succumbs to the allure of history. Instead of reaching for transparency, it affirms the natural meanings of words and uses visual devices to emphasize them, thus making the opacity of language yet more opaque. Evidently, it is trying to rehabilitate language rather than transcend it, a futile gesture if the language it tries to rehabilitate is impotent and sentimental. Such poetry is regressive, and indeed, many concrete poems have the quality of nostalgic lyrics. The difference between them and romantic poems by, say, Hölderlin or Poe is that they are minimal. Attention is focused on a few words or a single word. "l(a" by e. e. cummings uses one word exfoliated into four:

l(a
le
af
fa
ll
s)
one
l

iness.

Here the dictionary meaning of loneliness is reinforced by the image of a falling leaf. The image is both rhetorical and visual: the poem "falls" as one reads it. *One* and *i-ness* in the seventh and ninth lines further emphasize the theme. Aside from its brevity the poem is essentially a romantic lyric on the subject of mutability in the tradition of Poe's "Annabel Lee" or Tennyson's "In Memoriam." Cummings's object is not to discover new language but to reinvigorate a traditional one, the hidden agenda being to present the artist as physician with a charming bedside manner, even though the patient may be terminal.

The urge to revivify a language thought to be asthmatic and exhausted is interesting and significant and forms a prominent strain in concrete poetry. When Ernst Gomringer began to write concrete poems, he turned from German, which he had been using for conventional sonnets, to Spanish. Since he had been born in Bolivia, to revert to Spanish was to establish a link with his personal history. Concrete poetry, he explained in 1953, "must be closely bound up with the challenge of individual existence: with the individual's 'Life with language.' " Taking a similar position, Pierre Garnier argued in his *Manifesto* of 1962 that concrete poetry will make the word "holy again," while Ian Finlay, a Scots poet, announced that it "offers a tangible image of goodness and sanity." These are commendable sentiments. They are challenged, however, by the theory of Zdenek Barborka that poetry should transcend natural humanity, not to mention the announcement by the Japanese concrete poet Kitasono Katué that the insights of Zen have "driven [words] away to worthless rubbish."

Whatever the limitations of the romantic view of concrete poetry, many poems in this manner are extraordinarily effective. Décio Pignatari's "Beba" is derived from a found three-dimensional concrete poem—a Coca-Cola bottle. It uses its materials to satirize the spread of South American huxterism, which is derived, as Pignatari's readers understood perfectly, from Yankees to the north:

beba coca cola
babe cola
beba coca
babe cola caco
caco
cola
cloaca

Décio Pignatari, "Beba" (original in red type; 1957).

"Schützengräben" (the word means "trench"), by Ernst Jandl, is less political and more of a sound than a sight poem. It clearly owes something of its conception to the sound poetry of Kurt Schwitters. Its technique of exploded orthography is also evident in e. e. cummings's "l(a." It sputters with machine guns and whines with shell fire:

Ernst Jandl,
"Schützengräben" (1966).

schtzngrmm
schtzngrmm
t-t-t-t
t-t-t-t
grrrmmmmm
t-t-t-t
s——c——h
tzngrmm
tzngrmm
tzngrmm
grrrmmmmm
schtzn
schtzn
t-t-t-t
t-t-t-t
schtzngrmm
schtzngrmm
tssssssssssssssssssss
grrt
grrrrrt
grrrrrrrrrt
scht
scht
t-t-t-t-t-t-t-t-t-t
scht
tzngrmm
tzngrmm
t-t-t-t-t-t-t-t-t-t
scht
scht
scht
scht
scht
grrrrrrrrrrrrrrrrrrrrrrrrrrrr
t-tt

The result is a traditional lyric, but it is a good lyric. The sound effects powerfully intensify the meaning of the word. Perhaps they almost rehabilitate the word in this case.

One of the finest concrete poems in the romantic vein is Oswald Wiener's "Ich/Du." The poem offers a perfect minimalist visual statement about the relations between the *I* (*ich*) and the *thou* (*du*). The topic is philosophical, but it also relates to the idea of love, and the image is the merging at the center and then the separation of the *I* and the *thou*. The poem has the elegance and precision of a metaphysical poem like John Donne's "The Ecstasy":

du dich ich
du dich ich
dudichich
dich
ichdichdu
ich dich du
ich dich du

Oswald Wiener, "Ich/Du" (1965).

Without question the most famous American concrete poem is in the romantic tradition: Robert Indiana's *LOVE*. What could be more romantic than love? A much circulated version of this poem is an aluminum extrusion finished to mirror smoothness. It alludes by origin and finish to elements prominent in technological aesthetic. The letters that spell LOVE are stacked in two rows to create a shape with sculptural qualities—they visually objectify the idea of "support"—rather than a string of letters likc the letters on a printed page. However, the letters have serifs. The serifs come as a surprise. They allude to nineteenth-century printing and specifically to letters as counters that form words that are, in turn, to be read rather than seen. Complementing the serifs is a still more obvious allusion to the past: *LOVE*. No word is more saturated with tradition. Indiana's sculpture is intended to look modern, but it is really an elegy, a trip down nostalgia lane. It is modernist, not modern. It sold innumerable copies and has been canonized on a postage stamp.

The next most famous American concrete poem seeks transparency rather than affirmation of the past. The EXXON logo affirms the links, stressed so often by concrete poets, between art and technological culture. It is a complex marriage of form, color, and type. The advertising agency that produced it is said to have gone to extravagant expense to program a computer to create new words randomly until exactly the right word turned up. Finding a word that meant nothing was, in other words, a high-priority goal, and achieving the goal was anything but easy.

The search was complicated by several constraints. The first was that the word had to be be totally transparent. It could not mean anything in any language. Occasionally, brand names that sound fine in America and Europe have turned out to have embarrassing or indecent meanings in out-of-the-way languages.

The transparency requirement had additional constraints. The logo would be printed in Roman letters. It could not be pure symbol with no pretense of being a word, like (for example) the logo of TRW. It had to look like a word and be pronounceable and be reminiscent of the word it was replacing—ESSO. In the event, the word discovered by the agency computers began with an "E," was short, and had a double consonant in the middle. The double consonant proved the key to several problems. The two "X" 's are striking in themselves. Although double "X" is common in the Basque language, it is unknown in English. That is precisely its virtue. It ensures the transparency of the word. EXXON has no etymology, hence no history. It means only itself—"Exxon"—and that is exactly what it is supposed to mean. Indiana's *LOVE* is modernist; EXXON is modern.

The two "X" 's have another virtue. They are the basis of the visual expressiveness of the logo. The front leg of the first "X" descends below the bottom line formed by the other letters of the word. Crossed by another

EXXON logo. Photograph by Stavros Moschopoulos/Image Ray, 1987.

line, it forms the second "X." None of the letters has serifs. There is no sentiment here, although the echo of the "O" and of the double consonant in ESSO suggests a residual nostalgia. In many applications—for example, on gasoline station signs—the logo becomes sculptural. The letters appear in red on a white ground above a blue rectangle, the whole surrounded by a black frame with gently rounded corners.

In theory, concrete poetry emulates the transparency and precision of science. The EXXON sign seeks—and apparently achieves—these virtues. It is intelligible throughout most of the world without translation, and it means only itself. Typically, however, the words that are the raw material of a concrete poem draw it back toward traditional kinds of meaning in the manner of e. e. cummings's "l(a" and Robert Indiana's *LOVE*. It is curious that so many concrete poets yield to the allure of history while an advertising agency, which seeks only to communicate with the largest number of people for the maximum possible effect, uses the most radical and antihistorical strategies of concrete poetry. Obviously, there is money in transparency. Just as souvenir-collecting tourists were ahead of the Parisian artists in their appreciation of the Eiffel Tower, consumers of EXXON gasoline may know something that many concrete poets never learned.

2 1

OULIPO

Mathematical art is as old as the Greeks and as new as the shapes that emerge from computer images of the fractal equations of Benoit Mandelbrot. In his *History of Architecture* (1938), Banister Fletcher suggests that the spiral decorations ("volutes") on Ionic columns were drawn either by following mathematical formulas or by unwinding a cotton thread tied to a stylus from the mathematically curved surface of a whelk shell. Renaissance musicians experimented with a broad array of arbitrary systems of composition based on number, including systems based on rolling dice.

Mathematical elements are also common in literature. The Greeks and Romans thought of meter as "number," and Saint Augustine wrote an entire treatise, *De Musica* (*On Music*), to explain how the harmonies that meter generates embody the hidden perfection of divine numbers. The caesura in the first line of the *Aeneid*, Augustine explained, creates a metrical proportion of five and seven. Five squared is twenty-five. Seven is four plus three, and four squared plus three squared is also—amazing to say—twenty-five. Three, four, and five are also the sides of a Pythagorean triangle. The apparent imbalance of Virgil's line is revealed by mathematics to be a complex harmony.

In many ancient texts, most notably the Old Testament, elaborate numerological symbolism was discovered by the commentators. Because of Genesis, for example, seven is the number of temporal completion and of the temporal world in Judeo-Christian tradition, and eight is the number of eternity—a tradition honored by the tendency of ancient baptismal fonts, by which the convert enters into eternity, to be octagonal. Texts were also written that were explicitly numerological from the beginning. Dante's *Divine Comedy* is organized in threes in honor of the Trinity. An introductory canto of 136 lines is followed by three books of thirty-three cantos each written in "three's rhyme" (*terza rima*). The result is one hundred cantos leading the reader through three realms (hell, purgatory, heaven), of which the last contains ten heavens, the number of perfect completion.

More generally, "one" is the monad, the number of unity and of God. Three is the triad, the number of the Trinity. It is also the triangle, the most stable of figures and the number of dimensions, hence the number of space. Four is the number of the tetragrammaton, the four Hebrew letters that form the name of God that may not be pronounced, and the number of the four Gospels. It is also the number of proportion and the number of the four humors and the four compass directions. Five is the pentagon—among other things, the number of the body, which has five senses. Because seven is the number of the days of Creation, John Milton waits until Book VII of *Paradise Lost* to offer his own paraphrase of the biblical account.

Nine is three threes, the form of the emanations that pour outward into the world from the Nous, the Source of Being, hence the title of the great philosophical work of Plotinus—*Enniads* in Greek, *Nines* in English. It is also the number of the *Divine Comedy* considered as a progression of threes. Ten, the limiting number, is the sum of the first four numbers (1 + 2 + 3 + 4 = 10). The *Divine Comedy* is complete at one hundred cantos, which is also ten times ten.

Edmund Spenser's *Epithalamion*, which celebrates his marriage to Elizabeth Boyle, is based on astronomical and calendar symbolism. Analyzed numerically, its twenty-four stanzas record the twenty-four hours of the marriage day. The poem has 365 long lines, one for each day in the year. In the introduction to his religious epic *Davideis* (1656), Abraham Cowley explains the numbers of poetry by quoting the same biblical passage that D'Arcy Thompson quoted 250 years later in the majestic conclusion to his study of the mathematical basis of life: "The Scripture witnesses that the World was made in *Number*, *Weight*, and *Measure*, which are all qualities of a good Poem. The order and proportion of things is the true Musick of the World."

Quite aside from its mystical significance, number has an important effect on language. Since mathematics is one thing and natural language something else, mathematical patterning tends to distort natural language. The artist who decides to impose mathematical pattern on a literary work has two options. He can attempt to conceal the patterning and appear "natural." Or he can use the unnatural effects the patterning creates as positive and expressive elements.

Meter is a good example of the effect of number on language because it is so commonplace that it is usually accepted by readers of poetry without much thought. It is created by arranging words so their stressed and unstressed syllables form regular patterns, the most common pattern in English being the decasyllabic line, or iambic pentameter. Nobody arranges words that way in natural speech, but it is the standard speech form for Shakespearean dialogue. Since it is unnatural, its use conflicts with the impulse of dialogue to be realistic. Sometimes, especially in early plays like *Midsummer Night's Dream*, Shakespeare welcomes the artificiality of the form and emphasizes it further by rhyming his lines. At others, especially in the late plays, he deemphasizes meter by making his lines irregular and eliminating the pause that normally separates one line from the next.

Rhyme schemes and stanza forms also impose alien constraints on language. In correspondence with R. W. Dixon not published until 1935, Gerard Manley Hopkins proposed a curious mathematical interpretation of the sonnet based on the fact that its fourteen lines normally divide into eight ("octave") and six ("sestet"), with the larger units being subdivided respectively into four ("quatrain") and three ("tercet"):

It seems to me that this division is the real characteristic of the sonnet, and what is not marked off and moreover has not the octave divided again into quatrains is not to be called a sonnet. For the cipher in 14 is no mystery, and if one does not avail oneself of the opportunities which it affords, it is a pedantic encumbrance and not an advantage.

The equation of the best sonnet is:

$$(4 + 4) + (3 + 3) = 2 \cdot 4 + 2 \cdot 3 = 2\,(4 + 3) = 2 \cdot 7 = 14$$

Even if Hopkins's "cipher" is rejected as overingenious, the sonnet is an entirely unnatural form. The English variety requires pairs of rhyme words linked in the pattern a-b, a-b. Who would ever speak sentences in that way? The Italian sonnet is even more unnatural since it requires two sets of four rhyme words in the pattern a-b-b-a, a-b-b-a.

Romance stanza forms of the Middle Ages like the sestina, rondeau, and triolet use notoriously strict numerical formulas. The more complicated the stanzas become, the more they make poetry writing into a game. In the rondeau and the triolet, the play impulse that leads to the choice of stanza is often complemented by witty language. Wit, in turn, relates this sort of poetry to a tradition of "serious" comic poetry extending all the way from the comedies of Aristophanes to Wallace Stevens's twentieth-century mock epic "The Comedian as the Letter C."

The Japanese haiku is mathematically severe. It requires a sentence of seventeen syllables divided into three lines of five, seven, and five syllables respectively and in its traditional form must also refer to a month or season of the year. It is a fine example of poetry-by-number that has generated both serious and humorous verse in many languages. In English, two forms with even more severe constraints are the clerihew and the limerick. Both are used for humor. The master of the clerihew is Edmund C. (for Clerihew) Bentley, who wrote:

The Art of Biography
Is different from Geography.
Geography is about Maps.
But Biography is about Chaps.

Limericks range from mathematical to merely quizzical; from

'Tis a favorite project of mine
A new value of *Pi* to assign.
I would fix it at 3
For it's simpler, you see,
Than 3 point 1 4 1 5 9.

to

There was an old man from Nantucket
Who kept all his cash in a bucket;
But his daughter, named Nan,
Ran away with a man,
And as for the bucket, Nantucket.

The lipogram—a composition that elects to omit one or more letters—is another literary game. A famous ancient example cited by Georges Perec in his "History of the Lipogram" is the translation of the *Iliad* by Nestor of Laranda with "A" omitted from Book I, "B" from Book II, "C" from Book III, and so forth. Perec cites English lipograms that banish all vowels but one:

Idling I sit in this mild twilight dim
Whilst birds, in wild swift vigils, circling skim.

and

Lucullus snuffs up musk, Mundungus shuns.

To review the ancient and complex relations of number, game, and poem is to confront a paradox. All formulas for meters and stanzas distort language, and the strictest, most arbitrary formulas—for example, the formula of the limerick—produce writing that verges on nonsense and sometimes topples over the edge. Instead of avoiding such formulas, however, authors embrace them. In stricter formulas, the rigor seems to liberate the writer in two ways: from the need to make sense and from the tendency of language unfettered by formulas to run in dully predictable channels. A sign of this feeling of liberation is the playfulness that so often bubbles up in the poems. It is akin to the spirit of the "happenings" of the Cabaret Voltaire in Zurich. The difference is that Dada rejects all constraints, while forms like the limerick and lipogram use the most complex and arbitrary constraints to displace the patterns of grammar and syntax that normally control expression.

A question is surely in order. Why should intelligent artists accept formulas that force them to say things they might not otherwise say in ways that they most certainly would not use unless forced to by the formulas? Beyond the fact that they liberate, the appeal of the formulas is precisely that they force the writer out of predictable paths; that is, they encourage—in fact, they compel—discovery.

Since the 1950s there has been a revival of interest in formula poetry. The revival is associated with a group of writers, chiefly French, banded together under the name "Oulipo," which stands for "Ouvroir de Littérature Potentielle"—Workshop of Potential Literature. Oulipo was founded in 1963. Three of the founders of Oulipo were associated with the avant-garde of the first half of the century. Marcel Duchamp and Noël Arnaud had close relations with Dada, and Raymond Queneau with surrealism.

The continuity between Oulipo and the old avant-garde shows that in spite of its apparently arcane interests, Oulipo is in the mainstream of modern literature. Yet the basic impulse for Oulipo came from Albert-Marie Schmidt, a professor of French literature specializing in complex poetic forms like villanelle, rondeau, and lay. The elaborate formulas of fourteenth- and fifteenth-century poetry created a precedent for the formulas that Oulipo was interested in developing for twentieth-century literature.

The term "potential literature" comes from the fact that a single formula—for example, the formula for the rondeau—can be used for an unlimited number of specific rondeaux. The formula is a potential, and to analyze old formulas and make new ones is to establish a "workshop for potential literature." In addition to poets and scholars, Oulipo includes mathematicians, computer programmers, and experts in artificial intelligence. Evidently, it has staked out territory equally interesting to literature and to science.

Once the existence of literary formulas is recognized, it is natural to think about inventing new ones. So far, Oulipo inventions range from simple word-substitution formulas to formulas for increasing letter count as a composition proceeds to formulas involving repeated modifications of a text that depend on changes already made. The object of inventing the formulas is to see what happens to language when they are applied. Sometimes the formulas reveal interesting features of language or become expressive of themes inherent in the works that use them. At other times, the object seems to be to push language to the limits of intelligibility; or, in other words, to the point where it begins to disappear.

Typically the formulas specify procedures to be followed in the course of writing. Although they may seem to resemble the rules for creating a Dada poem formulated by Tristan Tzara, they are not formulas for random writing. Several resemble the formulas for achieving various results—for example, sorting words alphabetically—in a computer program. In honor of the latter, they can be called "algorithms" and the kind of poetry they produce, algorithmic poetry. Although the algorithms sometimes require the use of random choices, Oulipo rejects the idea of random composition. "There can be no doubt," says Jacques Bens, "about our aversion to the dice shaker."

In his "Foreword" to Warren F. Motte's *Oulipo: A Primer of Potential Literature* (1986), Noël Arnaud stresses the game aspect of Oulipo. Its playfulness contrasts with "the ponderous sobriety" of French structuralism. The source of the fun is the discovery of ever more clever formulas producing ever more amazing compositions. What is the difference between fun and madness? In several classic authors the play impulse verges on madness, and Oulipo celebrates the apparent madness of authors like Rabelais and Laurence Sterne.

In the course of inventing fractal geometry, Benoit Mandelbrot restored systems of mathematics to the light that had been called "monstrous" and locked away. In the same way, Oulipo has retrieved ancient and medieval literary forms labeled "frivolities" and "cretinous" and "pathological monstrosities" from the cabinets of philology. What is to be done about an author who uses a demanding formula but doesn't understand its implications? Oulipo explains the fact that authors retrieved from the cabinets of philology use Oulipo-like devices as "plagiarism by anticipation." For example, by using a trinitarian algorithm to organize every level of *The Divine Comedy*, Dante became a plagiarist by anticipation.

The madcap associationalism of Sterne's novel *Tristram Shandy* is another case of plagiarism by anticipation.

A proud boast of Oulipo is that it has produced the longest sonnet sequence ever written. Raymond Queneau's *Cent Mille Milliards des Poèmes—One Hundred Trillion Poems*—seems to consist of ten sonnets. Each is printed on a single page, and each page is cut into fourteen strips, one line to a strip. By turning the strips in random fashion the reader creates new sonnets. In fact, it is impossible to turn the strips in a way that does not produce a new sonnet. The number of sonnets possible is one hundred thousand billion—10^{14}, exactly the number promised by the title—"all structurally perfect and making sense."

The latter comment, by Martin Gardner in a review of Oulipo in the *Scientific American*, is a little chilling. One might wish (if not hope) that with enough flipping a determined reader might find at least one sonnet that did *not* make perfect sense.

Through science Oulipo has wonderfully multiplied human creativity. *Cent Mille Milliards des Poèmes* is truly a work of potential literature. A reader can read for a million years at five minutes per sonnet without reading the same sonnet twice. An anthology that would overwhelm the stacks of the New York Public Library has been close-packed into ten pages. There is a precedent in fractal mathematics for this kind of compression. By using algorithms that simulate reconstructions of images, it is possible to compress by a factor of 10,000 images that would require megabytes of space in a normal format. When reconstituted, the images are not quite what they were, but they are almost as good. Does this technique suggest the possibility of a fractal literature? For example, could fractal methods be used to create a reference book? Such a reference book would presumably be indistinguishable from a normal reference book, except that when its information was reconstituted, none of it would be quite accurate.

Different algorithms produce different literary genres. Ernest V. Wright's *Gadsby: A Story of Over 50,000 Words Without Using the Letter E* (1939) and Georges Perec's *La Disparition* (*The Disappearance*, 1969; reissued 1981) are novels that never use the letter *e*. They are gigantic lipograms. The absence of the *e* is a silence, a disappearance, a *blanc*. It signals the existence of a body of potential meaning that is unreachable unless the rules are changed. Without *e* there can be no "he" or "she" or "the" or "thee" in English, not to mention "be" or "love" or "fear," or, for that matter, "silence." Nor can there be "*le*" or "*elle*" or "*être*" in French.

No one can eat or mention eggs—*oeufs*—in *La Disparition.* This makes breakfast a silence. A character, Douglas Haig (pronounced "H'egg") Clifford, dies because during a performance of *Don Giovanni* he sings "*mi*," pronounced "me." Another character has six children, whose names begin with *a*, *a*, *i*, *o*, *u*, and *y*. *E* becomes a fearful presence through the consequences of its absence. An additional silence is imposed by numerology. *E* is the fifth letter in the alphabet. The novel has no fifth chapter. But silence must be overcome, and in spite of the absence of *e*, Perec sails as heroically as the speaker of "Un Coup de dès" over the ocean of the empty page—or, more properly, across a vast damp spot on a void folio.

At the beginning of the novel, a character, Anton Voyl, becomes obsessed with the figures in his oriental rug. He begins to fantasize. Eventually he disappears. Several friends, including a detective, gather at a château—Azincourt—to solve the mystery. There is an ancient curse on the château. They, too, begin to disappear. By the end the only character left is the author. His last sentence is "*La mort nous a dit la fin du roman*": Death has told us the story is over. The rest, of course, is silence.

La Disparition imitates a general social condition. The condition is exemplified in science by Darwin's inability to admit the aesthetic element in his style and in everyday experience by the Great Wall syndrome. Something is there but it cannot become conscious. Perec's characters frequently approach *e*, only to be baffled, frustrated, terrified, and, on occasion, obliterated. There are obvious Freudian overtones: repression occurs because certain things have been made unsayable. Perec's novel is a metaphor for a culture that has outrun its language so that its realities cannot be stated or even thought although they are continually experienced. It is a remarkable blend of madness and social analysis, a game that could not be played and a discovery that could not be communicated without the algorithm.

Martin Gardner, to whose discussion of Oulipo this survey is indebted, observes that the eleven most common letters in French spell the word *ulcération*. *Ulcérations* is the title of a book by Georges Perec published in 1974 by the Bibliothèque Oulipienne. In English the twelve most frequent letters, in order of decreasing frequency, are *etaoin shrdlu.* (SHRDLU, incidentally, is the name of a computer program using artificial intelligence to understand the position of a set of building blocks in an imaginary world.) These words formed the first and second columns respectively on the keyboards of old-style Linotype machines. When embossed on a Linotype slug they become a concrete poem of nearly perfect transparency.

An American poet, Jean Dunnington, has created a compound genre: the snowballing iceogram. Here, for example, is a dirge for a Scottie drowned at a picnic which includes an ironic digression on the indifference of nature to human sorrow:

O.
On!
One
Done.
Drone
Droned.
Drowned.

Dunnington's poem exemplifies the traditionalism that can be observed in concrete poems like "Avenidas" and *LOVE*. It remains loyal to natural language and meaning in spite of the pull of mathematics and pure sound.

Möbius poetry is poetry formed by writing two poems on a Möbius strip and reading it continuously. Such poetry can only be appreciated in the one-sided, one-edged world of the Möbius strip, since if the strip is severed, the result is two poems that are psychologically as well as physically scrambled.

In iterative poetry, a word or line is repeated. The musical composition known as a "round" is perfectly iterative because it repeats itself for as long as the singers keep singing. The refrains of ballads are iterative. Sometimes they are repeated without change and sometimes they change in relation to the stanza where they appear. In embedding, narrations are contained within narrations within narrations. A brilliant example is provided by the *Menelaiad*, which is a section of John Barth's

Lost in the Fun House (1968). In this novel the embedding of different narrative voices is indicated by alternating double and single quotation marks. The climax of the story follows the question "Who am I?" It is followed by a central moment, which is, appropriately, a silence: “ “ ‘ “ ‘ “ ‘ “   ” ’ ” ’ ” ’ ”. ”

An Oulipo algorithm attributed to Jean Lescure is called "S + 7." It requires substituting for each noun in a passage the noun that appears seven nouns ahead in a dictionary. With *Webster's Collegiate Dictionary* (1953) as the source, Hamlet's "What a piece of work is man" becomes "What a piecer of workaday is manager." This is not a bad thought but one that language would not have yielded up voluntarily. Counting "Pierian" as an adjectival noun, the algorithm "S + 14" produces: "What a Pierian of working capital is manchet." (*Manchet* means "loaf of bread.") That even this substitution is not entirely without meaning illustrates the stubborn refusal of language to disappear even when it is pushed to the brink. Different dictionaries and different decisions about what counts as a noun produce different results.

The Oulipo technique of transformation makes the question of meaning irrelevant and the disappearance of language explicit. In transformation, the words in two passages, chosen because of their striking differences, are replaced by their definitions. After several repetitions of the process, the two passages are shown to mean the same thing. Is the convergence accidental or a general condition of language? Jacques Robaud, professor of mathematics and member of Oulipo, concludes, "It has been conjectured that, according to this method, any two utterances in a language are always equivalent."

Perhaps the dictionary is circular. Perhaps everything eventually means everything else, which is almost the same as saying that nothing means anything. That has always been true, but people don't like to admit it because admitting it raises the possibility that the only reason language is not disappearing is that it was never there in the first place. This position capitulates to silence, and it is neither useful nor inevitable. The alternative is to use all available strategies to renew language or forge new languages. Dada used random strategies; Oulipo uses complex algorithms. Are the new languages more in harmony with the world than the old ones? Whatever the answer, they dramatize the problem of language. They also encourage appreciation for radical solutions—you might say for solutions that move from the strategies of *LOVE* to the strategies of EXXON.

There is an element in all poetry of surprise and originality, but surprise and originality are inseparable from obscurity. A truly original poem has a newness that makes it resist easy understanding. Oulipo algorithms distort language in surprising ways and produce effects that seem to many to be obscure and therefore poetic. *Are* they poetic? Or merely frivolous and fit for somebody's cabinet of curiosities?

Some of them will never escape the cabinet of curiosities. Some of them may become classics much as certain works of Dada have become classics. Are not algorithms as much a product of the human spirit as Kurt Schwitters's "W" poem? Perhaps more so in an absolute sense because they depend on conscious mental operations rather than on forms and instincts supplied by tradition.

22

ART AND IMITATION

If you begin with the ancient idea of art as imitation, it is clear that modern painting and sculpture are imitating several kinds of experience. You could say that the closest thing to a *real* thing—a Baconian object, hard and impermeable out there in the middle distance—is a rock. A rock can be the object of geological, chemical, environmental, mineralogical, physical, mechanical, or climatological studies, and each of these modes of analysis has its more specialized subdivisions. Each of the modes is valid, and in each of them an image of the rock is produced that is different from the images produced by the others. Where is the real rock?

Impressionism suggests that the rock varies according to the inner response of the beholder. Cubism goes further by imitating the structures of perception that create the thereness of the rock. The structures are spiritual forms, as Kandinsky would say. A process of disappearance is being imitated. Cubism imitates the fact that the more we know about the rock, the more ghostly it becomes. It tells us that the real is what we make it. This is imitation in the most primitive and simple form. It is imitation that might be called "making the obvious obvious."

Transparent poems are imitations that make the obvious obvious.

First of all, they confront the cultural reality of a world fundamentally discontinuous with its past. The obvious thing that the poetry of nothing imitates is the inability of natural language, as the secretion of thousands of years of history, to cope with the modern explosion of the new. The only thing as expressive as the poetry of nothing is silence, but silence is the shape of failure. In the old myths of the struggle of light with darkness, silence is the ally of darkness.

A second object of imitation of the poetry of nothing is the language that is emerging more or less spontaneously in response to the demands and opportunities of technological culture. When you buy a copy of *War and Peace*, you buy it because you know what it is and have decided you want it. When you subscribe to a newspaper, you do so because you know you do *not* know what the newspaper will say in the course of the coming weeks. The random quality of the front page of a modern newspaper is not accidental but essential. The root of *news* is *new*.

New is what is happening now, not what is being repeated. You may be able to anticipate a newspaper story about efforts to balance the national budget (though not the specifics of the story), but who can predict the thousands of stories about people and events totally beyond your knowing—the child falling into a well in Oklahoma, the production of human interferon from bacteria, a hurricane in Florida, the swearing in of the prime minister of Indonesia, rebellion in Chad, a world ice-skating record in Vladivostok? Like the aesthetic of the Great Wall, the aesthetic of the newspaper is invisible until seen apart from its utilitarian context, but it is powerful.

The Dada quality of newspapers is evident in other art forms created by high technology. Film and videotape can be mixed, spliced, and subjected to all manner of processing. Animation can be mingled with photography, colors can be modified, point of view can shift from microscopic to cosmic in an instant, and realism can mingle with surrealism and abstraction. These qualities were recognized at the very beginning of movies. The delightful and bizarre fantasies of the French filmmaker Méliès have the quality of magic shows. The same qualities are evident in films like *Blowup*, *Star Wars*, and *Sleeper*. The musical video, a phenomenon of the 1980s, routinely offers surrealism arising from the fusion of sound, image, and text.

Movies and television plays are produced by a series of "takes" bunched in groups determined on the basis of convenience and cost rather than the position of a specific "take" in the plot of the work being filmed. Takes from the beginning and end of a story are normally done at the same time if they use the same set because that strategy saves money. For the actor the drama unfolds as a jumble of scenes that are all from the same story but resemble so many unrelated fragments, like pieces of a jigsaw puzzle when they have just been poured out of the box. No matter how continuous a movie may seem to a viewer, it is a piecing together, a collage. This is not a matter of impression; it is a manner of the grammar of film, the way film is made. Viewers accept it with as little thought as the grammar they use to make their sentences.

As we have noted, in commercial television, an hour-long play is always interrupted by commercials. Other Dada-like events—demolition derbies, punk dress styles, singles bars, Led Zeppelin and Kiss, Atari computer emporia, supply-side economics—bubble up from the hot center of technological society like lava from a volcano. They are attributes of it, parts of its identity. Compared to such events, the poetry of nothing is conservative. This is as it should be. Art is not supposed to invent reality. It imitates what exists.

When the techniques of Dada and concrete poetry are used in commercial art or popular entertainment, no one questions them. They seem inevitable and entirely proper. It is only when they are named—that is, separated from their normal context and presented as what they are—that people look at them and feel surprised or shocked.

As with newspapers and television, so with language. Since the Renaissance, medicine has made new terms out of Greek and Latin words, thus attaching itself to history. The result has been an increasingly opaque

Duane Hansen, The Businessman *(1971). A striking example of superrealism.*

jargon, difficult for physicians, let alone patients. It has also spawned a swarm of nonsense words that *sound* like medical terms: "Pepsodent toothpaste contains irium." It is cumbersome and unexpressive when compared to languages that have abandoned history.

Acronyms are arbitrary formations, words without history since they have no etymologies. They are used in all advanced societies because they are as convenient and disposable as Kleenex. They form a temporary language—a language of discontinuity. Like the institutions they identify, they flourish and disappear like gnats in the spring: USIS, USIA, CIO, AAU, NRA, NAACP, CIA, ROM, ADFL, NASDAK, GAO, FICA, IBM, IDS, OPEC, IUD, COLA, DNA, NVKVD, AIDS, COBOL, BVD, SANE, SNOBOL, JDL, BIX, MANIAC, AEF, FEC, IRS, DOA, SHRDLU, HUD, DMV, DMZ, NASA, BMW.

Mathematics has also cut its moorings to history. It uses an alphabet that is part Greek (e.g. the "sigma" sign), part Roman (x, y, z), part Hindi via Arabic (1, 2, 3, 4, 5), and part arbitrary (the sign for infinity). The alphabet is complemented by operational symbols that mean only themselves, much as EXXON means only itself: plus (+), minus (−), "times" (×), "divide" (÷), "greater than" (>), "less than" (<), and the like.

Mathematics is a universal—hence a transparent—language. It has succeeded where Esperanto failed because technological society has no choice of whether or not to learn it. Not to learn it is to be swallowed in the deepest abyss of history, and no major culture is willing to have this happen. On the other hand, nobody has to learn Esperanto to survive. Therefore few people do.

Computer programs look like purest Dada to the uninitiated. One reason for this is that they are heavily acronymic. Probably they used acronyms originally because of the need for compression in the days of limited computer memories. That need is less urgent today, but the habit of acronyms remains as strong as ever in the computer world. The text that follows is the first part of a program in Lisp to control robots. Its aesthetic characteristics have intentionally been highlighted by giving it a title and printing it without explanation:

Meditations of a Robot

```
PUTPROP (QUOTE DEF)
  (QUOTE(LAMBDA(S A)
    (PUTPROP(CAR S)
      (CONS LAMBDA(CDR S))
      EXPR)))
  FEXPR)
PUTPROP(QUOTE DEF)
  (QUOTE(LAMBDA(S A)
    (PUTPROP(CAR S)
      (CONS LAMBDA(CDR S))
      FEXPR)))
  FEXPR)
```

In the light of this expression, the poetry of nothing begins to look less like an aberration than a prediction. By contemplating the obvious, Dada was able to anticipate languages for things that most citizens of the culture would not be able to see for many years. In this sense its imitations are examples of another ancient artistic function: prophecy.

Superrealism, as illustrated by the artwork of Duane Hansen, Richard Estes, and Don Eddy, produces images that seem more real than reality. They have an important message: we already live in a world in which the real is a manufactured product and in which animate and inanimate, human and artificial, are becoming indistinguishable; in which microbes are patented and cows have human genes and robots have replaced workers on the assembly line.

The two masses on either side of a geological fault grind against each other. Normally they release energy as a series of small shocks. Some of the energy, however, is not released. As it builds up, it eventually becomes so great that a major shock results. Enormous amounts of energy are released, and the landscape is permanently changed. The cultural equivalent of an earthquake occurred in Europe and America between 1900 and 1910. Artists have been trying to describe the new landscape ever since.

Scott Kim, Infinitely Spiraling Infinity *(1981). Computer poetry with a self-referential theme, using calligraphy with a mathematical basis.*

PART IV

ARTIFICIAL REALITY; OR, THE DISAPPEARANCE OF ART

Hair between lips, they all return
to their roots, in the blinding fireball
—NANNI BALESTRINI,
"*Hiroshima*"

23

COMPUTERS AND ART

Computer art began in the most propitious way possible, as a game. After the Second World War, large computers became sufficiently common to allow programmers and artists to play with them. Visual plotters had been developed to graph complex mathematical functions. The graphs often had powerful aesthetic appeal. They reminded people of pictures. Their complex curves, undulating, three-dimensional surfaces, and intricate patterns seemed to their delighted creators to be works of abstract art. Why not program computers to produce images intended from the beginning as works of art?

Why stop with drawings? Computers make noises and write texts. Why not make computer music and computer poetry? The closest thing to a formal proclamation of the coming of age of computer art was an exhibition held at Nash House in London from August 2 to October 20, 1968. An overview of the exhibition, entitled *Cybernetic Serendipity* and edited by Jasia Reichardt, was published the same year. In it Mark Dowson wrote:

> It is merely an historical accident that computers are largely used for mathematical calculation. Computers manipulate symbols which can represent words, shapes or musical notes as easily as numbers. Soon it will no longer be surprising to see a computer on the stage of Queen Elizabeth Hall—this actually happened in January 1968—interpreting and performing a piece of music before a fascinated audience.

The artists represented at Nash House began their careers as computer specialists rather than poets, musicians, or painters, and the association of computer artists with science and engineering—with places like the Jet Propulsion Laboratory, Bell Laboratories, and science and engineering departments of technical universities like Carnegie-Mellon and MIT—remains as striking today as it was in 1968. Most of them learned their craft on the job rather than through courses in fine arts. In this, they follow an honorable tradition of technological culture, going back at least to Gustave Eiffel. Among the artists featured in *Cybernetic Serendipity,* William Fetter worked at Boeing, and A. Michael Noll and Kenneth C. Knowlton at Bell Telephone Laboratories. Technically trained individuals remain prominent in the field. Melvin L. Prueitt, for example, works at the Los Alamos National Laboratory, Richard Voss at the Thomas J. Watson Laboratory of IBM, and James Blinn at the Jet Propulsion Laboratory at Cal Tech.

The proudest boast of the Nash House exhibition was that visitors were never sure whether they were admiring something made by a human being or by a computer. Around 1964 A. Michael Noll had begun creating semi-random images with an uncanny resemblance to Piet Mondrian's 1917 *Composition with Lines.* When Noll showed his pictures along with the Mondrian original he was surprised to find that most people were unable to identify the human artwork, and, in fact, 59 percent of the visitors to Nash House preferred the computer images because they seemed "less machinelike" and more "human."

The latter point is significant and applies to music as well as visual art. Much early computer music imitated the work of human musicians using traditional instruments—a Bach fugue, for example, played on a simulated organ. The music was amazing and delightful, but both the instrument and the performance seemed inhumanly perfect. Human music, it seems, is human not because it is precise but because it is imprecise. The imprecision comes in the form of innumerable small, randomly occurring variations from mathematical norms of resonance, timing, pitch, and the like. A definition of what it is to be human is thus an unexpected but provocative by-product of computer art: "To err is human." What distinguishes man from machine is the tendency to make mistakes. Once this truth is recognized, it is easy, in principle, to create machine simulations of humanity. If mistakes make a picture or a piece of music more "human" or "less machinelike" or "warmer," put them in the program. By examining closely the nature of human error, it should be possible, at least in theory, to create a mistake program that makes mistakes more inconsistently and hence more humanly than any human artist.

A computer program needs to be created and to this degree is the work of a human agent. Once it is finished, however, it becomes to a certain degree autonomous. It can produce thousands of different images or musical compositions or poems with minimum intervention. If the program calls for random changes in procedure, the products cannot be predicted from the program. Fractal images produced by Benoit Mandelbrot's equations can have self-similarity at different levels of magnification, but each level can be subtly and unpredictably different from all other levels. One of the fascinations of interactive computer fiction is that each "reading" of a work generates a different "plot." Some "plots" resemble other plots; on the other hand, some plots produce new and unexpected adventures. Interactive dialogues illustrate another kind of computer art. Because these dialogues develop in ways governed by the response of the human partner, they can objectify motives of the human partner.

The fact that many—perhaps most—of the pioneers of computer art were not trained as artists calls the position of the artist into question in exactly the same way that the powerful aesthetic elements of structures like the Crystal Palace, the Eiffel Tower, and the Brooklyn Bridge call it into question. Who is an artist?

The apparent autonomy of the programs that produce computer art raises a more fundamental question. If a computer program can create a series of Mondrian-like compositions that most viewers consider more human than Mondrian's original, should it not be considered an artist? Dada frequently produced art that resembles computer art, but Dada remained impure. In spite of their efforts to avoid it, Dada works often reveal the guiding hand of a maker. In this way they betray the lingering influence of the bourgeois habit of understanding art as a means of ego gratification. Computers complete the democratization of art begun by Dada, and as they do, they announce a change so fundamental in the idea of art that it can legitimately be called a disappearance.

2 4

FROM NASH HOUSE TO PSYCHIC SPACE

Several important truths about computer art had already become clear by the time the Nash House exhibition was mounted. In the first place, computer art was already extraordinarily various. The exhibition included graphs and mathematical functions; random-line drawings; imitations of abstract paintings; line representations of natural objects like flowers, seashells, and human figures; half-tone drawings; musical compositions; poems and story texts; visual poems resembling concrete poems; sculpture; and interactive compositions—that is, art works that react and change in the presence of viewers. In the second place, much computer art alludes to high technology as it makes use of the technology. This is true, for example, of graphs of mathematical equations offered as "paintings" and wire-frame drawings created by automated design programs. Much of the art is merely whimsical. None of it proclaims solemnly that it is "high art."

Three techniques that come naturally to computers tend to be used regularly, no matter what the style or the medium. The first is repetition (usually called recursion or iteration), which is the ability of a computer to repeat itself mindlessly until told to stop. If instructed to draw boxes, for example, a computer will happily draw picture after picture of exactly the same box for as long as the programmer wants it to. Simple repetition is numbing, not inspiring, and computer art supplements it with other techniques.

The second technique common in computer art is transformation. Transformation occurs when the rules governing a pattern are changed each time the pattern is repeated. The changes can be small, so they are not obvious until several repetitions have occurred, or they can be large and dramatic. In the early 1960s Charles Csuri, one of the few early computer artists with a background in art rather than science, produced a series of computer drawings of a girl's face. Each drawing changes slightly so that the series shows the change from girl to old woman. Csuri called the series *Transformation.*

The third common technique is randomness. This technique helped to make A. Michael Noll's Mondrian imitation seem more human than the Mondrian original. Randomness is common in twentieth-century art. We have already encountered it in the title of Mallarmé's "Un Coup de dés" and in Dada. It is regularly associated with liberation. As the art critic Meyer Schapiro points out in an essay in *Modern Art . . . Selected Papers* (1978): "Randomness as a new mode of composition, whether of simple geometric units or of sketchy brush strokes, has become an accepted sign of modernity, a token of freedom and ongoing bustling activity." Closely related to randomness, though carefully distanced from it by the authors of Oulipo, are mathematical algorithms and such intentional introductions of external influence into the creative process as John Cage's use of the *I ching* in musical composition.

Mathematical strategies obviously cannot be said to be caused by the use of computers, but computers make it easy to introduce algorithms, discontinuities, and randomness into every phase of the process of composition, and to do this in brutal or extremely subtle ways.

Randomness can be used to change one or more of the rules governing transformations and it can also be used to determine when the changes are to occur. Thus, for example, a computer can be told to draw a line within a rectangle, to stop after a random period of time if the line has not reached an edge of the rectangle, to bend the line at a randomly chosen angle, to repeat the process for a specified number of turns, and then to create an entirely new drawing by repeating the process. The result can be an impressive series of line drawings, each different from the next. In creating a poem, a computer can be given a list of nouns and told to select from the list at random whenever it needs a noun. The result will be a text—on occasion, a very interesting text. Computer music is highly developed and capable of an extraordinary range of effects. The word *aleatory* means "random" and is derived from the Latin word for "dice"—*aleae*. The use of randomness in computer music carries forward the experiments of many composers of the earlier twentieth century in aleatory music.

Between 1968 and the 1980s there were enormous improvements in operating speeds of computers and their memory capacities, in programming languages, and in the interfaces between operator and machine. Along with these improvements came high-resolution color graphics, high-fidelity stereo sound, color laser printers, high-level programming languages, intuitively transparent interfaces, and a variety of hardware devices adapted to the needs of the artist, including light pens, drawing boards, scanners, and synthesizers.

Computer-assisted design (CAD) programs allow complex shapes to be created in perspective, rotated so that hidden surfaces are revealed, and colored and shaded to create the effect of three-dimensional objects illuminated from a single light source. Ray-tracing programs allow precise control of light, including the accurate imaging of reflections on curved surfaces. VCR interfaces allow television images to be captured by computers and enhanced in various ways and combined with images generated by computer or recalled from the computer's memory. Through the use of fractals and other mathematical strategies, images have been created that could never have been seen without computers and that must therefore be considered unique forms of computer art. Fractals have also been used to produce lifelike imitations of natural scenes and objects, illustrated here by an imaginary landscape created by Richard Voss.

Computer images can also change. Interactive programs can link the changes to sound or motion or voice commands or any other variation in the physical environment of the image. Finally, imagery can be static or can move, and it can be isolated or it can be dynamically incorporated with music, nonmusical sound, words, and any number of other effects. As this happens art becomes holistic. At its farthest extension, it becomes a total environment controlling every sense rather than an isolated and thus objectified "experience" separate from other aspects of reality—as for example, the experience of a tourist from Knoxville looking at Rembrandt's *Portrait of a Lady with an Ostrich Fan* in the hushed and air-conditioned sanctuary of the National Gallery of Art in Washington.

Many of the techniques of the computer artist of the 1980s are so simple that they can be mastered by a clever computer-literate adolescent in a few hours. Others are extremely difficult, requiring the work of experts and the use of very fast supercomputers. The display screen used aboard Admiral Ackbar's spaceship in *The Return of the Jedi* (1983) is shown

Richard Voss, Fractal Landscape *(1982) Fractals used to create imitations of natural imagery.*

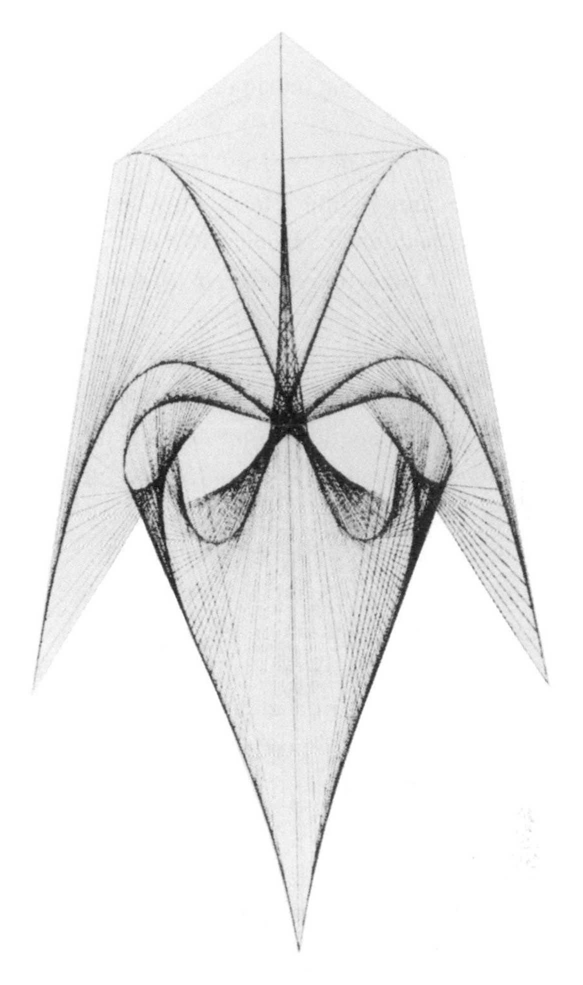

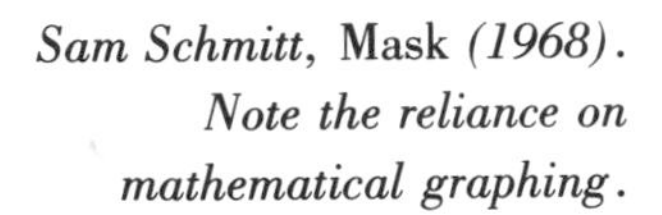

Sam Schmitt, Mask *(1968). Note the reliance on mathematical graphing.*

for thirty-eight seconds in the film. It took two Lucasfilm technicians four months to create those thirty-eight seconds.

Much computer art has a built-in urge to move. Being a product of technology and frequently having technology as its subject, it is drawn to F. T. Marinetti's ideal of an inherently kinetic art. *X-Wing*, by Art Durinski (1981), captures the feeling of motion, even though it is static.

Motion often becomes explicit in computer art. The ability of computer programs to generate a series of complex drawings, each a modification of its predecessor, makes computers ideal for animation. Initially, computer animation was jerky and far inferior in detail to the sort of hand animation made famous by Walt Disney movies like *Snow White* and *Fantasia*. Improvements in memory, speed, and screen resolution have closed the gap, and today color computer animation is so good—and so commonplace—that most people do not even realize when they are watching it.

The Walt Disney movie *Tron* (1982), based on a story by Steven Lisberger and Bonnie MacBird, was the first feature film to contain an extended sequence of computer animation. The sequence lasts fifteen minutes. In an article in *American Cinematographer* in August 1982, Bill Kroyer, one of *Tron*'s creators, is quoted as saying, "There was nothing done with the computers on *Tron* that could not have been done with conventional animation given 45 million dollars and one hundred years." The computer sequence created for *Tron* is stylized to take maximum advantage of what computers can do elegantly and to avoid what they do not do quite so well.

The images have a high-tech look that would be disconcerting in *Snow White* but is perfect for the high-tech theme of the film in which they appear. *Tron* begins when a hacker travels inside a computer. The viewer zooms over an idealized computer chip enlarged to the size of a city and later watches a thrilling race of high-tech "Light Cycles." The images are created to be viewed in rapid sequence. If one of them is viewed apart from the others, it can be admired, but it is best understood as a symbol of movement—something like a photograph that stops a baseball batter in midswing.

Computer image from Tron *(Disney Productions, 1982).*

Many of the adventures in *Tron* recall video arcade games. This is appropriate. The games that fill computer arcades all place great emphasis on motion. The earliest games—Pong, Aliens, and PacMan, for example—had only the most primitive graphics. The interest was in watching small, bright, player-controlled objects move around a monochrome screen, not in fancy imagery.

Later, the primitive games adopted color, and the little blobs of light began to become recognizable images of people and monsters and objects. But the fascination with motion remained a constant. In more sophisticated games—for example, race-car-driving and ski-slope games—the emphasis is especially strong because the images on the screen create the illusion that the player is in rapid, barely controllable motion.

The art of music is naturally kinetic; it can be appreciated only when it is in motion. Who can remain interested for very long in a single note? While kinetic images are viewed on a computer screen or through films of computer images, as in *Tron,* computer music is played by synthesizers or from recordings or on mechanized musical instruments.

A natural extension of the affinity of computer art with kinesis is reality simulation, in which the viewer is a creator of—or at least an active influence on—the reality being experienced. These games enact the plot of *Tron:* the player has in a sense been swallowed up by the computer.

Race-car-driving games are a familiar example of simulated reality. The playing area of a race-car game is usually an enclosed cockpit made to resemble a driver's seat. Sound effects simulate the growl of the engine revving up, the whine of acceleration, the screech of tires on a curve, the crunch of a fatal collision. Curtains isolate the driver from the everyday world. At first, the game is simply a game. As the speed of the car increases, more and more concentration is required to avoid destruction. Soon the concentration is so intense that the road is for all intents and purposes a real racetrack and the driver's cockpit is the seat of a real Maserati.

The original of all high-tech reality simulation was the Link Trainer, which was used long before the age of computers to train pilots in instrument flying. The pilot was placed in an enclosed cockpit, allowed only instruments, and given a series of problems by the trainer. Psychological realism was supplied by the readings of the instruments and by the realism of the problems encountered. It was reinforced by the pilot's isolation in the cockpit. The pilot of a Link Trainer was in a situation very much like that of the player of a race-car arcade game. The similarity is enhanced by the fact that in both cases the player knows that a fatal error will not cause death.

The techniques pioneered by the Link Trainer have important applications wherever difficult tasks involving high risk must be done rapidly and under considerable tension. Military situations frequently duplicate these conditions, and since the 1950s the military has underwritten the cost of many simulation programs. In these programs realism is vital. Pilots being trained to land on an aircraft carrier must have a realistic image of the carrier and the surrounding ocean, and the view must change as the angle of approach changes. To achieve this requires a powerful kinetic graphics program in color, capable of producing images that occupy a large percentage of the pilot's field of vision. Sound, motion, instrumentation, and the like must be coordinated in the program. The impulse is to create an environment so persuasive that when the pilot is concentrating, nothing will undercut the illusion that he is in a real airplane approaching the rolling, rain-slick deck of a real carrier.

A simulation called Shoot/No Shoot, developed by Bob Beckman while working for the Naval Investigative Service in 1981, is used to train police officers to make quick decisions about possibly life-threatening encounters. Realism is enhanced by the fact that each simulation program uses background scenery from the area where the training is being given. Another popular training system with the same emphasis on split-second decision-making is DxTr, made by Intelligence Images of San Diego to train emergency-room physicians.

The impulse to realism is never satisfied. Simnet places tank crews in exact simulations of an M-1 tank. The tank can sink in the mud, fire shells, and be hit. In the latter case, there is a powerful crashing sound and the "windows" that provide video simulations of the outside world go blank. Thomas Furness III, a designer of simulations systems, has been working since 1982 to create a "Super Cockpit" for the F-16 in which the pilot will have the option of blanking out the real world and relying entirely on images supplied by a "virtual world generator."

The relation of this and similar simulation programs to games is apparent in every popular computer magazine published today. One of the most popular games for personal computers is Flight Simulator, which is available in several versions. Flight Simulator provides a color-graphic display of the instrument panel of a single-engine plane and beyond it an image in color of the scenery being traversed and the airport being approached. Special disks are sold for different sections of the country. Each disk includes scenery, including natural and man-made hazards, unique to a specific area, together with the area's main airports. The view from the cockpit window changes as the altitude and heading of the plane change. Complex maneuvers, including loops, are possible, and novice pilots discover that their plane can stall and spin and even crash. The ancestry of Flight Simulator in pilot training programs is obvious. In like manner, Shoot/No Shoot is related to several war games requiring a fast draw, and anyone with a Commodore 64 or Apple II computer can buy a simulation game for basic surgery.

Here as elsewhere, it is almost impossible to draw a neat line between the serious and the playful in computer art. The ambivalence about what is play and what is serious recalls similar ambivalence in modern physics, in Dada, and in the work of Oulipo.

Reality simulation does not produce art one looks at in cool detachment like the Knoxville tourist contemplating Rembrandt's *Lady with an Ostrich Fan*. Nor is it like a movie, because movies are not interactive. The interactive aspect of reality simulation draws the viewer into the artwork and begins to do something else. By making the viewer participate in creating the artwork, it makes him to some degree the artist. Admittedly the reality simulation is not very powerful in arcade games, but it is there—a potential waiting to be realized fully in another context.

The work of Myron Krueger moves us from arcade games and pilot training to art. Krueger outlines his theories in *Artificial Reality* (1983). They began with what he calls "responsive environments" created at the University of Wisconsin, Madison, in 1969. The first environment was Glowflow. This was an empty rectangular room containing four tubes, each a different color, through which water containing phosphorescent particles was pumped.

The tubes provided the only light in the room. As people entered they activated floor pressure pads which, in turn, switched on lights at various points on the wall. The lights caused the phosphorescent liquid to glow and then gradually fade, creating patterns of multicolored illumination. Music composed using random techniques was also produced when the pads were activated. The program created random delays between the activation and the result. The result was an interactive environment in which there no longer seemed to be any relation between cause and effect.

As Krueger's ideas matured, he came to realize that his objective was the creation of what he came to call "artificial reality." In an exhibition called Psychic Space the floor had a maze pattern and the pressure plates responded with music that changed as the visitor moved. The space was intentionally "multidimensional" in the sense of creating questions that the visitor had to answer by exploration, but it was also a parable of the mystery of modern experience because "it was impossible to succeed, to actually solve the maze."

Krueger suggested that additional musical input could be derived from computer determination of the visitor's posture, so that walking through Psychic Space would become a kind of dance. As a matter of fact, the pressure plates are sufficient by themselves to cause dancing. Visitors to the installation of musical pressure-sensitive plates at Disney World's Epcot Center bounce from plate to plate, do impromptu arabesques, and repeatedly rock against a single plate as they discover the relation between their movements and the sounds that the space emits.

Further refinements were desirable. Krueger considered darkening the visitor with TV goggles instead of darkening the room. This would give the computer "absolute authority" over the visitor's field of vision. In another scenario, the environment might speak; perhaps its words would be coordinated with the visitor's footsteps "in a sort of mantra" or "a meaningless babble" which nevertheless would seem to "coax or warn or congratulate" the visitor. If the visitor spoke, the environment might respond with the sound of an earthquake or a train going through a tunnel, or it might turn the words into a musical composition.

Krueger's Videoplace used ten minicomputers. A video camera captured the image of each visitor, analyzed it, and produced visitor-specific music and imagery. Thus the visitor had no choice as to whether or not to assist in the creation of Videoplace. Videoplace had many scenarios and routines and used a different set for each visitor. In one routine, a life-size image of the visitor was projected on a screen. Beside the image was a small circle of green light which Krueger calls a "critter." If the visitor attempted to catch the critter, it would slide away.

A game between the visitor and the critter could ensue. Krueger explains that the critter is "a metaphor for one of the central dramas of our time: the encounter between humans and machines." Another routine permits the visitor to engage in finger painting. The visitor's hand becomes a hand on a video screen. Lights of different colors stream from the fingertips. In another routine a tiny door appears at the end of the room. As the visitor approaches it the video image shrinks in a way that recalls *Alice's Adventures in Wonderland.* On the other side of the door there is a blob of light. Whenever the visitor moves, the blob moves. It is the visitor's alter ego. As the visitor moves forward the blob is transformed into a fish . . .

Additional refinement will come, Krueger suggests, from an environment that is intelligent. The environment will then be something like an immobile robot that has ingested the visitor—an echo of *Tron.*

Krueger believes passionately in the liberating power of technology. Its achievements have "propelled us through a discontinuity." It has created artificial organs and artificial intelligence, and it has given man the power to control the most intimate evolutionary processes through genetic engineering. Man is "acquiring powers once reserved for the gods" and finds himself "not the final goal of evolution but its conscious agent." Krueger's "responsive environments" are a metaphor for the fact that there is no longer anything that can rationally be called "natural reality." Nature has disappeared. What remains is reality created or structured consciously by man for human purposes.

Krueger joined many other artists specializing in interactive artworks and artificial realities in the spring of 1985 in a spectacular exhibition entitled "The Immaterials" at the Pompidou Center in Paris. A technological optimism that Krueger must have approved of is evident in the explanation, quoted in the April 22, 1985, *Newsweek* of this exhibition by its planner, Jean-François Lyotard: "We wanted . . . to indicate that the world is not evolving toward greater clarity and simplicity, but rather toward a new degree of complexity in which the individual may feel very lost but in which he can in fact become more free."

2 5

SYNTHESIZERS AND NOCTURNES

Writing in 1968 in *Cybernetic Serendipity*, J. R. Pierce of Bell Telephone Laboratories complained that the computing demand of sound reproduction "strains the capacity of computers." Although computers have made remarkable progress since 1968, so has computer music. In an article written with Max V. Matthews for the February 1987 *Scientific American*, "The Computer as a Musical Instrument," Pierce notes that reproducing sounds with "a lush timbre" now requires up to twenty million operations per second for a composition with many instruments. The solution is to make the musical sound with a specialized sound-producing device called a synthesizer, connected to the computer through a standard interface called Midi.

Moderately effective music-generating stations can be created with a mid-range personal computer. A Mirage synthesizer, designed by the same engineers who created the Commodore 64 computer, sells for around $1,700. A Kurzweil, designed by Robert Moog, who is the father of the synthesizer, having created the first commercial model in 1970, costs around $10,000. A top-of-the-line rig like the Synclavier, produced by New England Digital, Inc., costs around $300,000, add-ons included.

Even as science delights in freedom, it wants predictability. It wants products that can be reproduced on demand and that are uniform in quality. When it observes the state of musical performance from this point of view it is disturbed. An orchestra is made up of violins, oboes, French horns, timpani, cellos, triangles, and assorted other instruments. These instruments are all different. Each instrument is also different from every other instrument of the same kind. One cello is considered outstanding, while another, produced by the same craftsman, is mediocre, although no one really knows why. You can say that the craftsman made mistakes when putting together the mediocre cello, but you might just as well say that the good cello was the mistake. In either case you have restated a point already made about the nature of humanity: "To err is human."

To make things worse, the method of playing each instrument must be learned laboriously, and each performer plays differently. Worse still, the same performer performs differently on different days of the week. Every live performance is different from every other live performance, which is to say that every live performance is a gamble. Even the most respected performers equipped with the best instruments can come up with a botch. Even a sloppy musician may one day blunder into a first-rate performance.

Recordings are one solution to this problem. In a studio a given piece of music can be performed several times. If one performance happens to be first-rate it can be used in its entirety. Otherwise, a synthetic piece can be created by patching together bits of several performances. In the past, the fidelity of recordings was limited, so that the best recording played on the best instruments failed to communicate the vitality of the original performance. This situation is improving as a result of digital recording techniques and optical disks, but it fails to confront another problem inherent in recordings. A recording pretends to be a live concert but it is something else. In the first place, it is static. In a concert, audiences can influence the performers by hissing or applauding or coughing or throwing roses onstage. To applaud or hiss a recording would be pointless. Like a book, a recording reduces the listener to the position of passive consumer. In the second place, as noted, a recording is often spliced-together bits of several performances rather than a single performance.

In the third place—and most fundamental—recorded music is sounds initially produced by vibrations of a needle in grooves or fluctuations of light reflected from pits on a disk. Converted to electric current, the fluctuations are processed in an amplifier and converted into sound by magnetically driven speakers. To listen to the sound they emit is not to hear instruments, only imitations at many removes from the real thing.

The alternative to a recording is an instrument that makes sound and that always sounds exactly the way it is supposed to sound and is played by a musician who always plays exactly the way she is supposed to play. Computer musicians have developed ways to approximate both of these goals, but for a long time the promise of computer music exceeded its ability to deliver. The sounds produced by electronic instruments lacked the richness of hand-crafted musical instruments, and electronic performance was only too obviously nonhuman.

Both points are illustrated by "Switched-on Bach," which was the first electronically performed classical piece to become popular. The Tempi recording was the work of Wendy Carlos on an early version of the Moog synthesizer. As represented on the Tempi disk, the organ is thin and whistlelike and its tempo is so regular that it instantly identifies itself as mechanically produced. A performance like this would have been fatal if it had been the work of a human performer, but it was acceptable and even appealing to those interested in electronic music. A review in *Electronic Musician* by Craig Anderton of the twenty greatest achievements in electronic music credits "Switched-on Bach" with stimulating a groundswell of interest in the subject. The Tempi disk was a best-seller when released and remains popular today. Why not? Bach's music depends heavily for its appeal on the sense of pattern and precise control of tempo and comparatively little on lush orchestration. Bach was also fond of the harpsichord, and one characteristic of that instrument is that it lacks the dynamics of the pianoforte. It *sounds* more like a computer instrument than a piano. Maybe "Switched-on Bach" is how Bach would have wanted to sound if he had had the technology.

Since "Switched-on Bach," computer sound has been made much richer by wave-form analysis. The wave form of an instrument is its characteristic sound including the overtones that give it timbre and warmth. Particularly important are the beginning (attack) and end (decay) of the wave form. A wave form is like a fingerprint. The wave forms of famous instruments—for example, a Stradivarius violin or a Yamaha guitar—can be sampled and stored. Later, they can be used in the synthetic orchestration of a musical composition.

The best samplers are very expensive, but once they have taken samples of a given instrument, the samples can be stored on computer disk and distributed to users. In 1987 I. M. Instruments was offering a disk with six guitars, three drums, and nineteen winds and strings, including harp glissando and organchord. In the same year E. C. T. SampleWare allowed purchasers to choose between an "orchestral" disk and a "rock" disk. With samplers, the automation of music is complete. Having completed the score of a composition, the composer selects the "instruments" to perform it from the computer's stock of samples, works out the orchestration, and then uses a synthesizer as a mini-orchestra to play the result. The finished piece can be retained in memory or printed out in musical notation or both.

This is more or less the same sequence of steps that composers have followed since the Renaissance, except for the use of wave-form samples for the instruments and a synthesizer to play the piece. Most computer music is more inventive than the scenario indicates, but even in the classic scenario, the result is different from a recording. The synthesizer does not pretend to be something it is not. When it plays, it is not a recording of anything; it is playing real instruments, although they need not be traditional ones like violins and flutes and woodwinds. The result is a performance of an electronic composition as "live" as a performance by the Berlin Philharmonic at Lincoln Center.

"Live" with one qualification. If the instruments in a classic computer composition are built from wave-form samples of real instruments, their sounds are analogous to sounds preserved in a recording. A performance using wave forms of traditional instruments is not quite "live" because it is unwilling to declare its independence from history—from the fact that musical compositions have traditionally been played by a limited number of instruments called violins and tubas and drums, and the related fact that some instruments, like Stradivarius violins, have gotten the reputation of being better than others.

An approach to music by Clifford Pickover of IBM illustrates the range of choices available to the computer musician. Pickover's "speech flakes" are computer-generated visual patterns resembling multicolored snowflakes with sixfold symmetry. In effect, they are pictures of various kinds of sound. Almost any data can be made into a speech flake—population data, the chemical composition of a compound, correlations among several types of data like employment, housing starts, and the value of the dollar, and more. The technique is new, but the principle of using nonmusical forms to create music was explored well before computers. Heitor Villa-Lobos used the skyline of New York and the contours of the Andes to define melodies on a conventional musical staff, a point illustrated by a poster for a book by Joseph Schillinger.

Using speech-flake technology, Charles Sweeney and John Holland of Michigan State University have created music from urine-analysis charts. Ivars Peterson explains in a July 19, 1987, article in *The Washington Post:* "A computer coupled with a Moog synthesizer translates spike heights into musical notes. A higher spike means a higher pitch." In this case the patient could be said to be playing himself, or perhaps the urine is playing itself.

Some artists continue to mix past and present. A 1987 University of Maryland Piano Festival featured a Bösendorf 290SE piano and an IBM personal computer. Under the tutelage of Morgan Cundiff, they played a Chopin nocturne. Cundiff was emphatically not creating a recording. His piano was giving a live performance of itself. The obvious analogy is to a player piano. However, a grand piano is emphatically not a player piano. Using a grand piano to create the sounds of a grand piano is a musical equivalent of the superrealism of Richard Estes and Duane Hansen. The product seems to be more real than reality itself.

The Maryland concert raises the problem of who is the performer of a computer composition. Cundiff did not perform the Chopin nocturne. He programmed the piano. To write a program is radically different from using the slippery confederacy of mind, muscle, adrenaline, talent, and training that go into a performance by a live musician. On the positive side, the Maryland performance had the consistency of a scientific product and can be reproduced in its original form any time the programs are

THE SKYLINE HAS ITS OWN MUSICAL PATTERN TRANSLATED FROM SILHOUETTE TO MUSIC NOTES WITH THE HELP OF

THE SCHILLINGER SYSTEM OF MUSICAL COMPOSITION

by JOSEPH SCHILLINGER

A scientific approach to writing music • Now available in Two Volumes

PRICE FOR THE SET: $30.00

Published by

CARL FISCHER, Inc.

Boston • Chicago • Dallas • Los Angeles • New York

Joseph Schillinger, New York Set to Music *(1946). The melody, based on the New York skyline, is by the Brazilian composer Heitor Villa-Lobos. Schillinger's system was used by several composers, among them George Gershwin, for* Porgy and Bess.

run. As long as neither the computer program nor the instrument is changed, each performance of Cundiff's nocturne will be exactly like the one before it.

This may be a flaw. Many music lovers consider variations in performance the essence of expressiveness and style. If the computer composer agrees, there are several ways to introduce them. Taking a cue from the response of Nash House visitors to A. Michael Noll's imitations of Mondrian, the composer can introduce random variations in the program. Since to err is human, the variations should make the composition seem more "warm" and "human"—perhaps more human than a human performance. An alternative strategy would be to analyze performance styles in the same way that wave forms are analyzed. With appropriate style analysis, the composer might be able to call not only for a Stradivarius violin or a Yamaha guitar but also for performance in the style of Yehudi Menuhin or Duke Ellington or Crystal Gayle. Since performers are human individually as well as in general and to err is human, suitable provision would have to be made for variations in each specific performance style.

A future Beethoven—or Michael Jackson or Madonna—might compose, for instance, a "sweatband quartet" for Yamaha guitar, Niagara Falls, Hoover vacuum cleaner, and wet sweatband. The composer can record it digitally, play back all or part or a single instrumental line, revise freely, print out a score, and create a synthesizer performance using specified instruments in the style of specified artists. At the same time, the composer can offer helpful medical suggestions to the donor of the sweatband. The result can be distributed on disk, downloaded to a thousand computer bulletin boards, mailed to the National Medical Library, and offered in concert at Queen Elizabeth Hall or the University of Maryland before an enchanted audience.

To review types of computer music is to be reminded of an important fact about the way technology enters culture and influences it. Some computer composers write music that uses synthesized organ pipe sounds, the wave forms of Stradivarius violins, and onstage Bösendorf grands in order to sound like traditional music. In this case the technology is being used to do more easily or efficiently or better what is already being done without it. This can be called "classic" use of the technology. The alternative is to use the capacities of the new technology to do previously impossible things, and this second use can be called "expressive."

The contrast between classic and expressive is found in many applications of new technology. Printing, for example, was invented to produce manuscripts more efficiently, and the first printed books were imitations of Gothic manuscripts and included hand-painted illuminations. In the same way, computer artists have made drawings of seashells and recumbent nudes, and writers have regarded computers as nothing more than superefficient typewriters. Because of their contrasting attitudes toward the past, classic and expressive can be understood as equivalents in the area of technology use to modernist and modern styles in art and literature.

It should be added that the distinction between classic and expressive is provisional because whenever a truly new technology appears, it subverts all efforts to use it in a classic way. Thus even as they try to be classic, users of a new technology contribute to the obsolescence of the classic. For example, although Gutenberg tried to make his famous Bible look as much like a manuscript as possible and even provided for hand-illuminated capitals, it was a printed book. What it demonstrated in spite of Gutenberg—and what alert observers throughout Europe immediately understood—was that the age of manuscripts was over. Within fifty years after Gutenberg's Bible, printing had spread everywhere in Europe and the making of fancy manuscripts was an anachronism. In twenty more years, the Reformation had brought into existence a new phenomenon—the cheap, mass-produced pamphlet-book.

A truly new technology refuses to stay classic. Even if it was first created for a classic function, it eventually becomes expressive and reshapes the function. If printing is a good example, another is the automobile, which began as an improved version of a horse-drawn carriage. The success of the automobile created so many new conditions that society had to be reshaped to accommodate them. In spite of the best of early intentions, within a few years after its commercial introduction the automobile ceased to be classic and became expressive. In the same way, in spite of many classic applications, computer art has an innate tendency to be kinetic and to exhibit itself on video screens rather than in pictures hung on walls.

"Switched-on Bach" seems like a textbook example of a classic use of a new technology. However, the reason for its popularity was not that it sounded like Bach. There were innumerable recordings available when it was issued that sounded a great deal more like Bach. The reason for its popularity was that it sounded like electronic music. It was sold for the whistlelike purity of its sound and the robotlike precision of its tempo. It was expressive in spite of itself. It did not offer a rehash of the past but something new. Some people detested it—as some people will dislike any new art form—but many found it delightful.

26

THE UNIVERSAL INSTRUMENT

J. R. Pierce wrote in 1968, "The computer can become the universal musical instrument." In 1987 he insisted with equal fervor that computer music offers "a virtually limitless universe of sounds through which composers and performers can express their thoughts and feelings." Pierce arrived at this conclusion when he recognized that what is called an "instrument" in traditional musical terminology is, in electronic terms, a "circuit" connecting devices that create sounds called unit generators. Inside a computer, a Stradivarius violin is a complicated digital pattern. So is a speech flake expressed as a sound. So is the sound of dishes breaking. So is any sound.

Pierce was familiar with Max Matthews's program called BTL that made circuits "of almost any degree of complexity." He adopted Matthews's terminology. Each circuit thus became an "instrument," and a group of "instruments"—no matter what instruments—became an "orchestra." Since the number of possible circuits is effectively unlimited, he could offer composers an almost infinite number of instruments, most of which had never been used (or even heard) before. By corollary, orchestras could be assembled, at least in theory, transcending in size and variety anything known since the beginning of music. Why, then, limit the orchestra to instruments that were familiar to Brahms?

Composers from Vivaldi to Gershwin to Edgar Varèse have experimented with the use of instruments to create "unnatural" (i.e., noninstrumental) sounds—wind, church bells, thunderstorms, locomotives. These experiments have frequently carried over into the use of noise-emitting objects as instruments—anvils, typewriters, automobile horns, surf. The result is called *musique concrète.* The computer musician can digitalize any form of noise and incorporate it into a composition. The technique can bring reality and history together in new combinations: the Holland Tunnel at rush hour played in the style of Yo Yo Ma.

"Pro Midi Studio," offered to computer musicians in the mid-1980s by Soundscapes, Inc., has an echo mode whose program "checks the distance from middle C . . . and starts a new track transposed the calculated distance from middle C each time a new key is pressed. This can create huge textual washes à la Philip Glass. . . . When used tastefully, the results . . . are fabulously hypnotic."

In a review of this program in *Amiga World* (July 1987), Ben and Jean Means, two knowledgeable students of computer sound, observe that its sampling device permits making instruments of "your dog, your kids, your pots and pans. Dishwashers, lawnmowers, phone calls, bouncing balls, kitty cats, baseball bats. . . . In [our] latest effort a chorus of breaking bottles, ringing crystal glasses, thumping cardboard tubes and a motorcycle accompanies the guitars, bass and synths." Since wave forms can be modified once recorded, "You can draw sounds from scratch, sounds never before heard by the human ear." And, of course, there is the synesthetic music of speech flakes.

Following grants in the 1960s to Yale computer musicians by the National Science Foundation, J. R. Pierce observed, "If good art can embody a valid contribution to good science . . . then art can validly share not only the fruits of science, but the support which society so rightly gives to science." Computer music deserves support from NSF because it is in the vanguard of a techno-cultural transition. Computer musicians who use their universal instruments to create facsimiles of traditional sound are in a certain sense reactionaries. Instead of making "a valid contribution to good science" they are making facsimiles for people who want to believe that, after all, nothing has changed.

At MIT's Experimental Music Studio (EMS), Pierce's concept of the "universal instrument" has developed in two ways. Under the guidance of the studio's director, Barry Vercoe, a "synthetic performer" is being developed. The goal is to create an artificial performer capable of replacing any human performer during a musical performance so efficiently that the other live performers "cannot tell the difference." A second objective is to create "hyperinstruments" that instantaneously convert any human agent into a composer-musician of titanic power. Since both the synthetic performer and hyperinstruments eliminate the traditional skills of the musician from music in the same way that computer drawing programs can eliminate the artist, they point in a clear direction—to the disappearance of musical art.

One of the most important world centers for computer music is the Institut de Recherche et Coordination Acoustique/Musique (IRCAM) located at the Pompidou Center in Paris. The institute is directed by Pierre Boulez, who refers to his six years as director of the New York Philharmonic as "a big parenthesis in my life." One of the achievements of the institute is the creation of the "4X" audio processor, which is capable of handling up to five hundred independent sound sources. Since it opened, Boulez has encouraged computer music and has himself composed several brilliant works that intermingle live and computer performers. "Répons," written in 1981, is for six instrumental soloists, chamber orchestra, and digital processors. The processors transform the music created by the human performers and play it back as a "response." Displacement is central to the composition. As the music is displaced in tempo, pitch, and apparent location of the instruments, the soloists are displaced spatially around the circumference of the concert hall. The instruments are cymbalon, xylophone, glockenspiel (doubling the xylophone), harp, two pianos, and Yamaha DX-7 synthesizer (doubling piano).

Another important composition associated with the Institut de Recherche is *Valis*, a media opera by Tod Machover based on a novel by the late science fiction writer Philip K. Dick. Machover studied with Boulez at IRCAM and is now at the MIT Experimental Music Studio. When *Valis* was performed in Paris in December 1987, it featured (in addition to live performers) two hundred video screens, continuous computer transformation of the music created by the performers, and manipulations of the human voice that required breaking sentences into fragments and recombining them so that they modulated in and out of intelligible pronunciation of the words. At the end, the speaking voice of an actor was transformed electronically into song so that the actor accompanied himself.

In spite of the impressive achievements of *Valis,* Machover was dissatisfied. Commenting to Stewart Brand, whose *Media Lab* (1987) traces many of the most exciting artistic developments taking place at MIT, he imagines a symphony that creates itself: "You might just push a button and watch it behave, or it might be a performance system—you could interact with this structure at any level of detail you wanted."

An emphasis on the MIT Music Studio and the Institut de Recherche is justified in view of the importance of these two centers to current developments in computer music, but it is misleading if it suggests that they dominate the computer music scene. No list can do justice to the number of active centers or the richness of computer music since 1970. To cite only one example, James Terry studied electronic music and phenomenology at the University of Illinois and then spent two years at Bell Telephone Laboratories writing computer music. During this period he created six compositions using the full range of expressive computer technology. "Noise Study" begins in a traffic tunnel. Terry wanted to draw on characteristic twentieth-century aural experience: the random sounds generated by a dynamic technological culture.

His first step was to create "an 'instrument' that would generate bands of noise, with appropriate controls over the parameters, whose evolution seemed most appropriate to the sonorities I had heard." Since these sonorities had to be "framed" in order to inform an audience that they should be listened to—i.e., should be understood as aesthetic statements—Terry established a large-scale structure with note values determined "by various methods of random number selection." He recorded the piece on tape and mixed the resulting sounds with sounds of the same tape played at half the normal speed.

Mike McNabb's "Dreamsong" (1977–78) is a mixture of sounds ranging from the voice of a soprano to Dylan Thomas reading a poem. James Moorer's "Perfect Days" (1976) mixes a human voice with a flute in the reading of a poem: "The result is both astounding and beautiful: a synthetic voice speaking comprehensible words, yet retaining the breathy, silvery sound of a flute."

The music of the League of Automatic Music Composers is produced by "the dynamic conflict and cooperation of four automatic composition programs" running together. "On Being Invisible" (ca. 1975) by David Rosenbloom begins with random sounds, selects "structural landmarks" determined by "brain signals of the performer" (who becomes the performer through the act of listening), and produces musical compositions based on maximizing given signals from the performer's (that is, the listener's) brain.

If computers are good at producing patterns, they are also good at distorting patterns. Random number selection is an important element of James Terry's method for the same reason that it is important in visual art and poetry. It breaks up forms inherited from the past. In an article written well before the advent of computer music entitled "Music and Chance," Alfred Schreiber remarks: "There have been a number of composers . . . who introduce chance not as a point of comparison for criticism of style, but as a starting point or music-generating principle. It has become customary to group together such compositional procedures using the term 'aleatory.' "

As noted earlier, *aleae* is the Latin word for dice. Mozart entitled one of his compositions "Musical Dice Play" (*Würfelspiel* in German). John Cage creates "pure, unbridled acoustic chances" by the use of the *I ching*, but he considers the *I ching* "temperamental" rather than random in the sense of indifferent and he objects to pure randomness as strongly as the authors of Oulipo. Yannis Xenakis, another aleatory composer, has used "Poisson's law of the distribution of random events" as his guide to creating revelatory sound. Computers are not necessarily better than the *I ching* and Poisson's law at producing randomness, but they simplify the incorporation of randomness into the deep structure of compositions.

Herbert Brün's "Soniferous Loops" (1964) is an early example of the technique. It is described by Brün as a series of "random sequential choices channeled and filtered under control of form-generating restrictive rules." In 1967 John Cage joined Lejaren Hiller, a pioneer computer musician at the University of Illinois, to create a computer-determined aleatory composition, *HPSCHD*, scored for seven harpsichords. Appropriately, the basis for the score is Mozart's musical *Würfelspiel*. Cage approves especially of the fact that aleatory techniques allow the composer to "renounce control," a reminder of the desire of the modern artist to sever the still powerful bourgeois link between art and ego.

27

WHITE SOUND, WHITE LANGUAGE

In nature, all forms are moments that join forms below with forms above in a sequence that seems to extend infinitely in both directions. Nothing that is fractalized has a final "form." As the sequence is followed, another phenomenon is evident. There are niggling small irregularities and surprising large irregularities and discontinuities. An art that incorporates randomness can imitate both the niggling irregularities of nature and the large discontinuities. It becomes, you might say, a natural art or an art of nature.

Here we enter a most interesting area: differentiation among art forms by medium.

Random frequencies of light produce something that is technically not a color, but that is perceived as a color: white. Tiny bubbles in a polar bear's fur create the illusion that the fur has a white pigment because they scatter all frequencies of light equally. White pigment is used regularly by artists, and the white paper of a watercolor pad can be as important to the final picture as the watercolors. In spite of not being a color, white is a positive resource for the painter, not a mere background. White pattern is less positive. Wallpaper might be considered a metaphor for white pattern because it is a continuity that prevents a surface from being an emptiness, although, it should be added, a white wall is white, not empty.

No positive quality like "white" is evident in completely random sound or language.

However, the properties of sound and language are, themselves, different, and the concept of randomness helps to define the difference. In acoustics, sound composed of random frequencies is called white sound. It is perceived as a hiss. This hiss is of interest because it has unique properties. A tape recording of white sound refuses to change pitch when slowed down or speeded up. Since the sound consists of frequencies equally distributed, changes in the speed of the tape have no effect on it.

White sound ought to be a resource like "the color white." Actually, it is a kind of silence because no message can be extracted from it, but it is a positive silence, a silence that is a presence. In this respect it is like wallpaper. People dislike absolute silence as much as visual emptiness. White sound is sometimes added to recordings and tapes because silence seems less "empty" when it is used. A composition that approaches the purity of white sound needs no score. It has thrown off the constraints of ego to become a sensual expression of the world beyond the mind.

What applies to sound does not seem to apply to the medium of speech. Speech is a phonetic code the essence of which is organization—self-defining differences among elements. "White language" should be like white sound, but it is perceived as nonsense rather than a presence offering shelter from emptiness. In written language, a randomly created alphabet would not be recognizable as an alphabet. A text in this language would be an absolute silence because it would never reveal the fact that it is a text. It might be redeemed by the images of the letters if they were interesting. Regarded as images, the letters might elicit favorable judg-

Joseph Stiegler, Transformations *(1968). A computer-generated random alphabet might look like this.*

ments. However, regarded as letters of an alphabet, they would be dismissed or create resentment because they would seem to invite understanding and frustrate it simultaneously.

Because music in which randomness is a structural principal rejects historicism, it is disorienting if "framed"—that is, presented in a way that demands it be listened to. It threatens to absorb the listener in an oblivion like the condition induced by prolonged sensory deprivation, or, perhaps, like the surrender of the self to half-seen, gliding beautiful things glimpsed in a moment of revery. Yannis Xenakis seems to be hinting at something like this in his statement that his listener should be "gripped, and drawn willy-nilly into the circle of notes, without any special training being necessary. The sensuous shock must be as palpable as that of hearing thunder or looking into a bottomless chasm."

Of course, you do not have to choose between absolute disorder and absolute order. In Mozart's "Dice Play" the disorder acts as a dash of seasoning to the order. There are patterns of randomness as well as simple randomness, and chaos theory—which is related to fractal geometry—explores situations in which order is born from disorder and disorder from order. If totally random sound is white sound, perfectly orderly sound ought to be called black sound. Somewhere between white and black there is brown sound, which is pattern built on a foundation of randomness like a Turkish rug spread over a bottomless chasm.

28

THINGS SEEN AND UNSEEN

Computer art arouses little of the hostility that greets computer music and poetry. Part of the explanation is that many computer images have a high degree of order. Most familiar are the chunky bar charts and pie charts produced by spreadsheet programs. Publishing programs generate sophisticated type fonts and complex page formatting. Television advertising has made computer color animation commonplace.

Long ago, Sam Schmitt of Princeton produced images representing cubic functions. Most people find such images striking and many consider them beautiful. Melvin L. Prueitt of the Los Alamos Laboratories has explored graphlike imagery more fully than any other well-known computer artist. *Gold Wing* (1981) announces by its title that it is to be regarded as art rather than mathematics. In addition to being beautiful

Sam Schmitt, Cubic Interpolation 1 *(1968). An image that insists on being recognized as art but announces clearly its derivation from mathematics.*

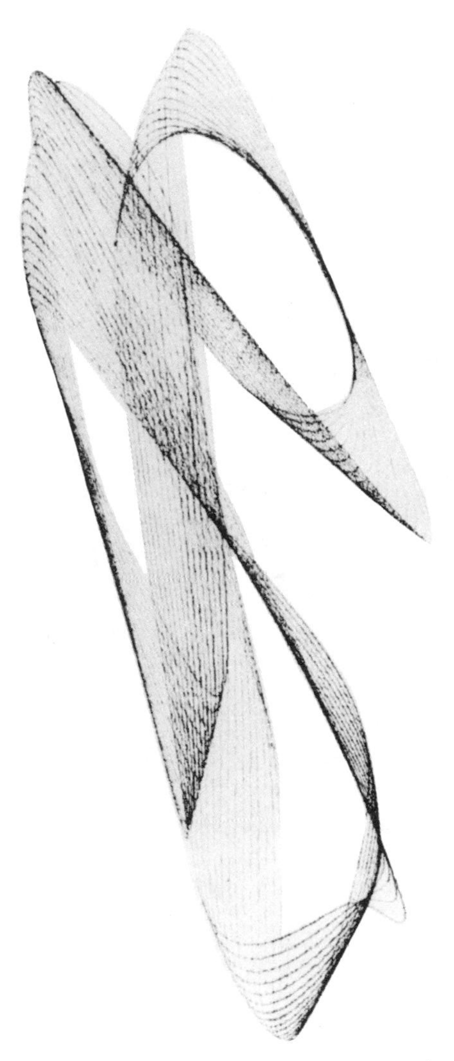

it touches a responsive chord (see color picture 6). Anyone who used a Spirograph as a child has been fascinated with similar forms.

Computer-assisted design programs (CADs) are more representational. Frequently they begin with wire-frame images that are later converted to solid three-dimensional representations. These images can be rotated, set in motion, and continuously modified. Engineering design programs tend to produce images that seem practical rather than aesthetic. However, when the object drawn is striking, the drawing often has powerful aesthetic appeal. Solid modeling often produces even more striking images, a point illustrated by Mike Newman's drawing of the American space shuttle. Images can also be created by computer programs of objects that could not otherwise be seen, such as hypercubes and the doughlike

Mike Newman, Space Shuttle *(1982). An image produced by solid modeling techniques of computer-aided design (CAD).*

four-dimensional fractal shapes created by Alan Norton at IBM's Watson Research Center.

Ronald Resch, a professor of computer science at the University of Utah, is known in the United States for his design of the mouth of the alien spaceship that swallows the starship *Enterprise* in the first *Star Trek.* He is famous in Canada for another kind of computer art: a twenty-six-foot, two-and-one-half-ton egg created in 1976 with James Blinn and Robert McDermott for the town of Vegreville in Alberta, Canada. The egg is made of 2,732 precisely designed and brilliantly colored aluminum tiles supported (like the Statue of Liberty) by an internal frame. Other artists draw inspiration from programs created to depict molecules in three dimensions (Nelson Max) or to make three-dimensional architectural renderings (Michael Collery).

The picture types considered thus far all illustrate the classic phase of computer art. They aim at utility or, as in the case of Melvin Prueitt's *Gold Wing,* are extensions of programs that were originally intended only to produce useful images. The images are schematic or representational, even when what is being represented is nonvisual, like temperature fluctuations or airflow. An indistinct and shifting boundary separates these classic images from images that are expressive. The expressive images are not engineering drawings treated as "found art" or extensions of techniques used to produce engineering drawings. Instead, they are pictures intended as pictures, even though they often allude to engineering techniques.

Melvin Prueitt's *Sunflowers Watching the Sun* takes a giant step toward being expressive. Prueitt's sunflowers are stylized, as is the setting in which they are placed, but they are recognizable flowers, not graphs, and they lean like real flowers to a common source of light. David Em, who does his work at the Jet Propulsion Laboratory in California, is one of the most brilliant of today's computer artists. *Bill* is a composition that combines a stone backdrop and a brilliantly modeled egg in a haunting and surrealistic combination of textures, colors, and spaces (see illustration 7 in color section).

Melvin Prueitt, Sunflowers Watching the Sun *(1983).*

Mike Newman, Chambered Nautilus *(1981). Here the mathematical basis of natural forms merges easily and naturally with the mathematical bias of much computer art.*

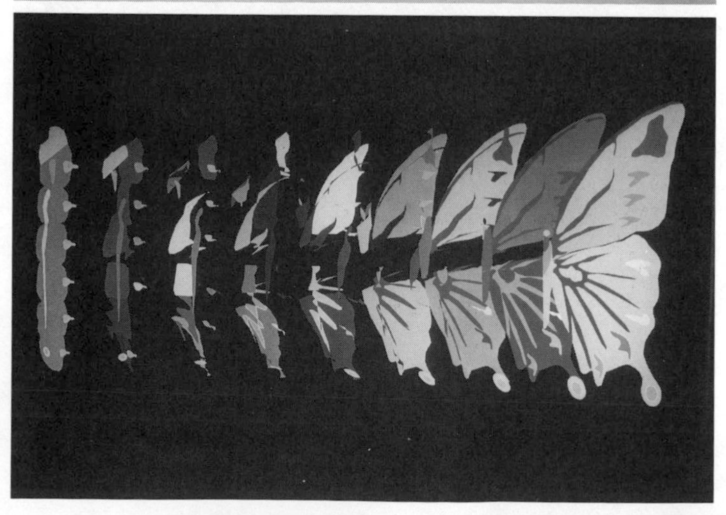

Mike Newman, Butterfly *(1981). Another instance of the impulse toward kinesis.*
Opposite: Cockpit Man *(1968). Computer-generated studies of the human body in constrained positions created for the Boeing Corporation.*

Kerry Strand's *The Snail* (1968) nicely illustrates the way that an expressive composition can allude to classic techniques. The shell of the chambered nautilus has a precisely defined spiral shape. Strand represents it as a lacy and delicate mathematical curve. For obvious reasons, the chambered nautilus haunts artists who are drawn to the mathematical aspects of computer art. Another theme that has fascinated computer artists is metamorphosis, illustrated here by a butterfly created in the 1980s by Mike Newman.

The human form has always appealed equally to scientists and artists. The point is equally obvious in the musculature of Michelangelo's nudes and in the illustrations in the great atlas of anatomy published by the physician Andreas Vesalius in the mid-sixteenth century. Barry Arnalt of Boeing Aviation became interested in the human figure for very practical reasons. In the 1960s he wrote a program for drawing a pilot in various postures to facilitate cockpit design. His drawings are therefore classic rather than expressive. However, like the drawings in Vesalius, they protest against this classification. A page of Arnalt figures has a little of the feel of a page of sketches from the notebook of a sixteenth-century master. Arnalt's figures are, in turn, the obvious ancestors of a sequence of lighthearted and entirely expressive sketches made by Intergraph that are used as section headings in this book.

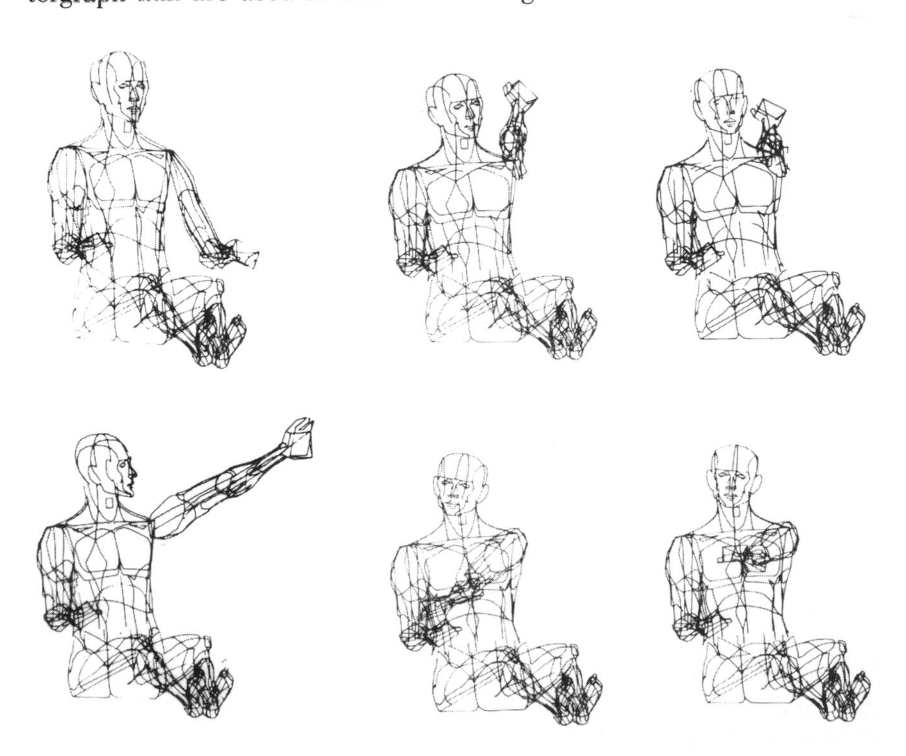

Image enhancement and manipulation are classic techniques useful in a broad range of applications from interpreting satellite images of the surfaces of distant planets to criminal identification. Using digitalized photographic images, any artist who fancies the idea can redraw and recolor the *Mona Lisa* and can combine her image, redrawn or not, with other images and image fragments. The same techniques can be used to create deformations of photographs.

Redrawing the *Mona Lisa* has been a hobby of artists ever since Marcel Duchamp affixed a moustache to her upper lip, and deformations of the human face have been common in modern art since Picasso. More interesting from an artistic point of view are techniques that use computer programs to distort systematically. These techniques are related to the manual techniques used by D'Arcy Thompson in *On Growth and Form* to produce deformations of animals long before computers came onto the scene. Computers, however, can move faster and deform in many more ways. *Monroe in the Net* is created by breaking lines in a grid. Deformations based on mapping one form on another become powerful social commentary in the images of the head of President John Kennedy produced by the Computer Technology Group (CTG) of Japan in 1967–68.

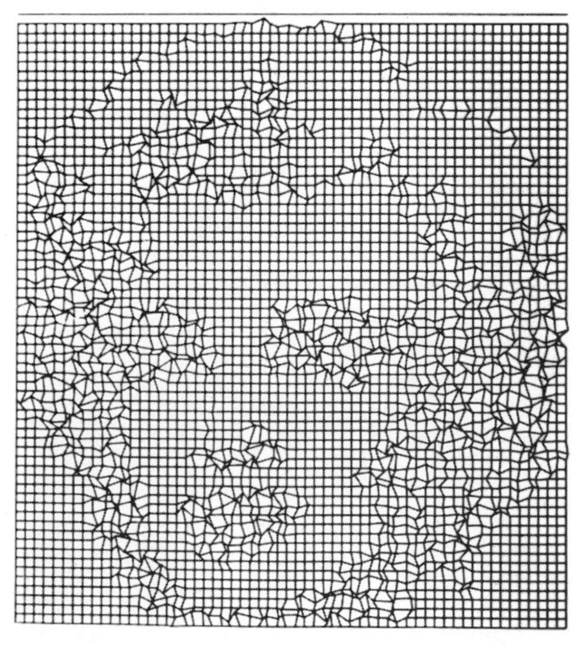

Computer Technology Group (CTG) of Tokyo, Japan, Monroe in the Net *(1968). Computer art has, from the beginning, explored deformations and surrealistic effects. A kinship can be recognized between this interest and the geometrically deformed fish and animals in D'Arcy Thompson's* Of Growth and Form.

Computer Technology Group of Tokyo, Shot Kennedy No. 1 *(1968). An example of deformation combined with powerful social commentary.*

A happier example of computer art is provided by the work of Harold Cohen, a British computer artist. Cohen uses a Logo turtle to draw the outlines of his images and then paints in the colors manually. The technique combines the sophistication of the human artist with the charming clumsiness of the mechanical turtle that executes the artist's wishes. The aesthetic result is a Chagall-like playfulness far removed from the obsessive precision of more typical computer art. Appropriately, one of Cohen's best murals adorns the wall of the National Children's Museum in Washington, D.C.

Robert Parslow and Michael Pitteway of Brunel University have taken the opposite path. They have sought to eliminate the human artist entirely from the production of the artwork while retaining the sense that each work is unique. Like Cohen, they are repelled by obsessive precision. Their program is intended "to escape from the pretty geometrical figures of conventional computer artwork" and gain "a new freedom" of aesthetic expression. A drawing from the program suggests that a certain amount of freedom was achieved. The drawing is almost random, but its confinement hints at the presence of a concept. For those who know how it was created, it poses the question of whether freedom is an absolute or a complexity of control so dense it cannot be perceived.

Parslow and Pitteway did not produce a drawing but a program capable of making endless drawings in which "every drawing is original" because "the computer never repeats itself." Does this suggest that the artist may one day become superfluous and thus disappear? In a 1960 *Encounter* article entitled "Inventing the Future," Dennis Gabor wrote, "I sincerely hope that machines *will* never replace the creative artist, but in good conscience, I cannot say that they never *could*."

Robert Mallary, a sculptor, not only foresees the disappearance of the artist but celebrates it. Writing in *Art Forum* in 1969, he describes the process in lucid detail. He and his students had created a modest sculpture program called Tran2. The exercise enabled Mallary to formulate criteria for the next generation of programs, many of which are commonplace in the CAD programs of the 1980s.

The experience of the jump from first- to second-generation programs led Mallary to predict that computer-assisted sculpture would evolve rapidly. He described the process in six stages. By stage two the computer is "indispensable" because it does calculations beyond the capacity of the human artist. Among its contributions at this stage is "randomness . . . particularly important . . . precisely because it is *not* a fixed and absolute thing." By stage four the computer is making decisions that the artist cannot anticipate. The result is "a continuing redefinition of the role of the sculptor." Mallary considers the possibility that by stage four the computer will be so fully attuned to its artist that "posthumous production" of works "in the manner of" the deceased artist will be possible. At stage five "the machine will have capabilities so superlative

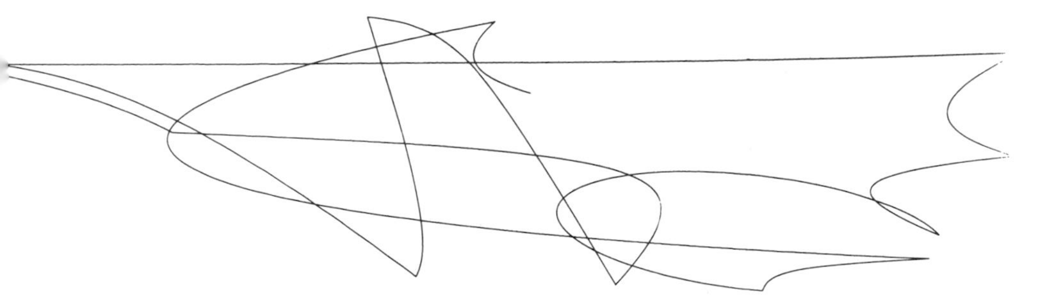

Robert Parslow and Michael Pitteway, Random Drawing *(1968). Computer art as pure abstraction. The program that produced this drawing can produce any number of drawings, each different and, in that sense, "original."*

that the sculptor, like a child, can only get in the way," but the sculptor can still pull out the plug. By stage six "the sculptor, if the word has not long since . . . been discarded entirely, will probably not be able to pull out the plug."

29
ADVENTURE

Although computers are tolerable musicians and brilliant artists, they seem to many to be dullards when it comes to language. Louis Milic, a pioneer in computer poetry, observed in 1970 that the objective of those who want computers to use language must be "more modest than that of the computer musicians, who have some respectable compositions to their credit, and the producers of computer graphics, whose beautiful arrangements of lines and colors adorn a number of walls in good artistic company."

In spite of their limitations, computers are probably used more generally in writing than in any of the other arts. This is because in writing they have an obvious and powerful classical function. They seem to most writers to be extraordinarily effective combinations of the yellow pads on which much writing used to be initiated and the typewriters on which later drafts and final copies were made.

Nothing has changed from the days of the typewriter, or has it? There is a lot of controversy about this among writers and teachers of writing. The consensus seems to be that a lot has changed but that the changes are under the surface.

Take format. If you ask the traditional writer how long his book is, he will look at a pile of smudged, interlineated typescript and reply, "About 350 pages" or perhaps "Around 100,000 words." If you ask a computer author the same question, he will say he isn't sure, or maybe he will check the disk directory and observe that the manuscript is 420K. If he has an antic turn of mind, he may say it is about a quarter of a mile long. Computer text exists as a long scroll in the computer's memory. It appears by screenfuls to the author, and until it is printed out, the readiest measure of length is the number of kilobytes it occupies in memory, which can be translated metaphorically into the length of the imaginary scroll. The point is not entirely facetious. A conventional manuscript exists as pages. It is easy to compare page 20, say, with page 276. This is possible but less convenient for computer texts. The thrust of computer writing is continuous movement ("scrolling") from one screen to the next.

Take form. Does the opportunity that the computer offers for endless revision change the writing process? Certainly. Knowing that revision is easy encourages logorrhea in some writers. It encourages others to take chances. It may paralyze writers who are neurotic about revision. There is always time for one more run-through. Nitpick follows nitpick until the original work threatens to dissolve and has to be started again from scratch.

Now take something that goes a little deeper into the issue. Take process. Educational literature of the 1980s is filled with discussions of computer-assisted writing. The subject can be conveniently sampled in articles appearing in the February 1988 issue of *Academic Computing*: "Redefining the Book," "Word Processing and Writing Process," "A Networked Classroom," and "Writing Technology and Secondary Orality." "A Networked Classroom" stresses the fact that with computers writing no longer needs to be an isolated activity. A writer who uses a local area network can communicate actively with others while engaged in the process of writing. The experience can be psychologically reinforcing. It also introduces a new concept: writing as a communal process. A networked classroom is a "discourse community."

When the community begins to collaborate, writing becomes anonymous, which is to say that the author begins to fade away. The point is not at all farfetched. Many traditional publications are anonymous, and such publications adapt well to computers. Who is the author of the telephone book? The same question can be asked about catalogues, large reference works, collections of statistics, repair manuals, and the like. Indeed, memos become more and more anonymous as they circulate through a large office for suggestions and criticisms.

Networked classrooms are a classic use of computer technology. When we move to the expressive use of computers in writing, the place of the author becomes still more problematic. Computer novels, for example, are seldom the work of a single author. They are more like folk ballads. After they have been circulating for a few years, they may have several layers of input from many tinkerers.

Like ballads and folktales, they also exist in variants so that the versions downloaded from different bulletin boards are likely to be different. Public-domain computer programs go through a similar process of modification and augmentation, and comments contributed to computer conferences are likely to be so augmented, revised, explicated, and annotated that they eventually become almost unrecognizable to their originators. Walter J. Ong suggests in *Orality and Literacy* (1982) that computers encourage a return to many of the mind habits of oral literature. As they return to modes of communal art that predate literacy, they move beyond the concern with author and ownership that characterizes the culture of the book.

What about content? We have already noticed that computer art has an innate urge to become holistic. Computer writing is no exception. Beginning with the bar charts and pie charts, computers encourage writers to be visual. Scanners allow pictures to be "captured," and desktop publishing programs and laser printers allow the result to be printed.

Once pictures become an integral part of the writing process, the writer's use of language changes. A real estate brochure "written" by computer might have one picture showing a house from the street, another showing the floor plan, and another showing the recreation room during a party. The text accompanying this brochure will let the pictures speak mostly for themselves and concentrate on things that cannot be shown in pictures—financing details, proximity to schools, shops, and entertainment, and the like. The "writer" of the brochure is as much a visual composer as a composer of words. In this respect, computers carry forward the movement to reclaim the visual element in language that is so important a motive for concrete poetry.

Imagine an art history "text" on optical disk including color reproductions of two hundred key pictures. The reproductions are in high resolution. They can be seen entire or segments can be examined in magnified detail. A painting of the Virgin Mary now in a museum can be superimposed on a photograph of the church wall where it originally hung. The Elgin marbles in the British Museum can be viewed attached to the remains of the Parthenon in Athens. A repair manual for a transport plane includes diagrams of the parts and the parts of parts, together with verbal comment and suggestions about each maintenance task. The suggestions are updated as improved parts replace old ones and as maintenance crews make discoveries about their job and pass the information along to other crews.

The resources made available to the writer by computers force a rethinking of what "writing" is. Since the same person who is creating the text is selecting and placing the images, they are part of the "writing." They are no longer illustrations interpolated by an outsider. In general, writing that includes images is less dense and less continuous than language as developed for the presentation of complex ideas without images. Vocabulary can also be simpler. Remarks and captions replace long descriptive paragraphs. The result can be more informal, more colloquial, more user-friendly than the traditional language of books.

Where, you may say, does computer writing allow for the brilliant and powerful effects of "pure writing" as illustrated by Chaucer's *Troilus and Criseyde* or Melville's chapter on "The Whiteness of the Whale" in *Moby-Dick* or the stream-of-consciousness passages in Joyce's *Ulysses* or Alice Walker's *The Color Purple*? The answer—if we consider computer writing in its expressive rather than its classic phase—is "nowhere." Pure writing is different from the often communal and imageoriented manner of composing that seems to be natural for computer authors. In its expressive phase, computer writing does not try to duplicate the past but to do new things. Why not? Eighteenth-century authors did not produce medieval romances; they produced novels. The masterpieces of computer writing—if and when they appear—will be different from the masterpieces created by "pure writers."

Then there is the effect of computers on the literature of the past.

One of the most-discussed developments of the mid-1980s in the use of computers in education is hypertext. Gregory Crane of Harvard outlines the subject in an excellent article in *Academic Computing* entitled "Redefining the Book." Hypertext is basically a database program, and other programs are capable of emulating many of its abilities. To keep things simple, all such programs will be referred to generically as hypertext.

As the title of Crane's article suggests, when used to present standard literary works, hypertext is an extension of the annotated books that most people will recall from courses in Shakespeare. There are three important differences. In the first place, a great deal more information can be incorporated into hypertext than in a printed book. In the second place, the information can include complex images—maps, schematic drawings, engravings, and photographs. In the third place, although footnotes are usually placed visibly at the bottom of the page or, perhaps, on a facing page in a textbook, hypertext is normally invisible until invoked. When invoked, it can provide not only the sort of information presented in footnotes, but also documentation of the notes, notes on notes, sources behind sources, images and details of images, dictionary definitions and the sentences on which the definitions are based, bibliographies of specifics, and bibliographies of bibliographies. Hypertext can provide these tools in a structured manner as part of a planned survey of a topic, or it can allow the user to range freely, assisted by the remarkable speed and flexibility of access it allows to the available data.

The virtue of hypertext as applied to literary studies is that it allows many different resources to be brought to bear on a given work. A hypertext version of Shakespeare's *The Tempest,* for example, would include the text itself. It would also include a list of variants in the text. *The Tempest* doesn't have many, so the list would be short. However, a list of variants for *Hamlet* or *King Lear* would be long and complicated. For *Hamlet* there would also be sample pages of "good" and "bad" quartos and of the text as it appears in the First Folio of 1623.

There would certainly be discussion of the date of *The Tempest,* relating its theme to the beginning of English colonization of the New World. The texts of two narratives of the shipwreck of a boat called the *Sea Adventure* in 1609 in Bermuda and the letter by William Strachey describing the shipwreck should also be available because one or more of them was a source for the play. A map of the New World showing where the *Sea Adventure* was wrecked would be useful. Another source that should be included is the essay "Of Cannibals," by Michel de Montaigne, in the English translation published by John Florio in 1603.

Staging is important in *The Tempest.* A hypertext *Tempest* would include a map of London showing the location of the theatres, as well as drawings of the stages of the Globe and Blackfriars as scholars currently imagine them, and drawings of possible blockings of key scenes.

The hypertext would include explanations of difficult words like the explanations in the footnotes of conventional texts. It might also include explanations of points of English Renaissance grammar unfamiliar to modern readers. Magic is important in *The Tempest,* and there is an elaborate masque featuring classical deities. Hypertext should include an explanation of the Renaissance idea of magic, including the distinction between white and black magic, and an explanation of what a masque is. Also useful would be drawings of seventeenth-century masque costumes and of famous productions of *The Tempest* showing how various actors and costume designers have imagined Caliban and Ariel. There should also be . . .

But enough! The possibilities are endless. The limiting factor is storage, and optical disks have already created storage capacities for personal computers rivaling those of mainframes of the 1970s. The hypertext *Tempest* is a teacher's dream of what a student reading *The Tempest* should have.

What does hypertext do for—or to—*The Tempest*? Unfortunately, the answer is not as simple as it might seem to be in the abstract. The clear implication of hypertext is that *The Tempest* is not a literary work to be enjoyed but a heap of facts to be memorized or a puzzle to be solved or a mystery to be explained.

If you imagine a reader using hypertext, you have to imagine a constant movement from text to glossary to grammatical comment to classical dictionary to Bermuda map to textual variants to drawing of Ariel to text to *Sea Adventure* narrative to . . .

If the items just listed were read in the order given, they would form a collage. They are not linked because Shakespeare put them together that way in *The Tempest* but because of the interest or curiosity of the reader or the demands of a teacher. The process of "reading" hypertext is therefore different from what is normally meant by the phrase "I read the play." With hypertext, the process is interactive and discontinuous—almost the opposite of reading *The Tempest* continuously from beginning to middle to end or seeing it performed.

The reader must constantly decide whether to read a sentence in the play or invoke the drawing of the Elizabethan theatre to find out how the shipwreck might have been staged or turn to the glossary to find out what "unstanched wench" means or read somebody's explanation of why the play begins with a shipwreck, or dip into the description of the wreck of the *Sea Adventure*, or what have you. The sum of the fragments chosen becomes the text the reader has actually read. Each "reading" produces a new text—a metatext—made of bits and pieces of Shakespeare interspersed with bits and pieces from other sources.

When "read" in this way, the play tends to disappear into the hypertext like water in a sponge. Admittedly, computers did not create this situation. The disappearance of the literary text began in the nineteenth century with the theory that literature should be studied as science studies nature. However, computers make the disappearance of the text irreversible by plunging the reader into an information swamp from which there is no escape other than turning the computer off and pulling out a paperback copy of the play—or, better yet, attending a performance, since the performance does not have footnotes.

Hypertext makes clear a fact that was often noticed before it appeared on the scene. We are coming to the end of the culture of the book. Books are still produced and read in prodigious numbers, and they will continue to be as far into the future as one can imagine. However, they do not command the center of the cultural stage. Modern culture is taking shapes that are more various and more complicated than the book-centered culture it is succeeding.

At the same time, books in the sense of novels and plays and short stories are becoming harder to appreciate in the old way. Long before hypertext, modern culture developed the habit of equipping literary classics with prefaces, appendices, footnotes, collections of critical essays, and bibliographies of additional readings. These imply that without the aids supplied, the books in question are in danger of being unintelligible—that is, invisible. Literature is also and increasingly the object of quasi-scientific analysis that examines issues remote from the interests and experiences of literate readers. Computers encourage this approach. An article by Estelle Irizarry, a literary scholar, describes the use of Lotus 1–2–3 to make a "graphic configuration for analysis of novels." Irizarry applies the system to *Cauce sin rio* (1962) by the Puerto Rican novelist Enrique Laguerre. Her graphs are impressive. They reveal patterns the reader of the novel would otherwise ignore, and that is both the point and the problem. The point of the analysis is not to clarify what the reader of *Cauce sin rio* is likely to find obscure but to reveal things about which the reader is likely to be totally indifferent.

Only a short distance separates hypertext from the interactive computer novel. An interactive novel is associative. Like hypertext it works because of a series of links among its parts. Some of these links are permanent. Others change in the course of each reading.

One of the earliest important interactive novels was *Adventure*, and this novel is still available in many flavors from computer bulletin boards. *Adventure* began in the 1960s as an exercise in artificial intelligence programming by Willie Crowther and Don Woods of Stanford. It operated by means of a dialogue between the computer, which asked questions and made comments, and the player, who typed answers that drew further comments from the computer. The game motif has remained constant in all versions.

The "reader" must decide how to proceed and what to use to cope with the situations encountered. To open a door, for example, the reader has to have decided to pick up a key that was lying on the floor of a room entered earlier. Many of the problems are extremely difficult and require reasoning, guesswork, and occasional fudging in the form of queries to other players. Over the years, as *Adventure* became more sophisticated, the dialogue was enriched. Descriptive passages were improved, and some versions had sound and graphics added.

The story begins near a small house. You (the reader) are invited to enter. There are several objects, including the key just mentioned. The reader leaves the house. One turn leads to the woods where you can wander without progress until you are fed up and quit the game. Another turn leads down a riverbed to a grate. If you have picked up the key in the house, the grate can be opened. If not, you have to return to the house. You now pass through a series of underground chambers. In one there is a bird. A dwarf may appear menacingly from the side of the chamber. You continue, descending to a passage blocked by an enormous serpent. The bird can vanquish the serpent, but there is no explanation from the computer that you need the bird. How do you know? Well, one way is to recall that birds and snakes are mortal enemies. Another way is to ask a friend who has already gotten past the snake. Suppose you don't have the bird. There is an alternative path, but it is blocked by a chasm. How to get across?

Some readers proceed stolidly forward. Some ask for help. Others get angry. While I watched in horror, one player became so frustrated that she killed the bird. *Adventure,* in other words, is self-referential. You do not read about an adventure, you have an adventure. There are dangers and triumphs. A high point in the game is Don Woods's justly celebrated description of "The Breathtaking View": "Far below you is an active volcano, from which great gouts of molten lava come surging out, cascading back down into the depths. The glowing rock fills the farthest reaches of the cavern with a blood-red glare, giving everything an eerie, macabre appearance. . . . The walls are hot to the touch, and the thundering of the volcano drowns all other sounds. . . . A dark, foreboding passage exits to the south."

The setting of *Adventure* recalls spelunking expeditions that Willie Crowther took in Colossal Cave in Kentucky. Everything is underground. The plot, which Don Woods improved, is influenced by Tolkien's *Lord of the Rings*. But can you call it a plot when each "adventure" is different? In an article on "Interactive Fiction as Literature" in the May 1987 *Byte Magazine,* Mary Ann Buckles calculates that there are 187 billion trillion different possible ways of experiencing just the "battery maze" of *Adventure,* a number that recalls the *Cent Mille Milliards des Poèmes* of Oulipo.

Adventure is a pioneer work. More modern interactive novels make dazzling use of animation, color, and combinations of music, synthesized voice, and sound effects like waterfalls, hoofbeats, and clashing swords. Most important, perhaps, the challenges become more devious and the responses of the computer become more like human responses.

Among the best-sellers of 1986 were *The Zork Trilogy,* which has become a classic since the release of *Zork I* in 1979; *The Hitchhiker's*

Guide to the Galaxy, which is derived from a science fiction novel by Douglas Adams; Tolkien's *The Hobbit;* and a mystery story entitled *Where in the World Is Carmen San Diego?* that comes with a *World Almanac of Facts* needed to follow Carmen's meanderings across Europe. In the mid-1980s a new genre of fiction became popular—cyberpunk. The foremost classic of cyberpunk is William Gibson's *Neuromancer.* Dr. Timothy Leary, guru of the 1960s, has formed a company to turn *Neuromancer* into an interactive game or, to use the terms supplied by the advanced hype, "a mind play or performance book." The reader of this work will apparently create the screens or "acts" as the game proceeds. Leary also claims to have rights to works by Thomas Pynchon and William Burroughs.

Possibly the most brilliant of interactive novels is *Mindwheel,* by Robert Pinsky. A poet as well as an interactive novelist, Pinsky traces *Mindwheel* to Edmund Spenser's *Faerie Queene* and Lewis Carroll's *Through the Looking-Glass.* The story is based on the premise that powerful minds leave psychic impresses behind them. The minds in the novel belong to Bobby Clemson, an assassinated rock star; The Generalissimo, a dictator executed for war crimes; The Poet; and Dr. Eva Fein, a female Einstein. The reader must penetrate these minds.

The novel thus becomes a psycho-voyage through alien consciousness. The goal is to reach the Wheel of Wisdom, which is held in the mind of a prehuman creature who invented the lever and the rhythmic chant but who also "resembles Jackie Robinson and Yvor Winters." Pinsky's references extend from Dante to Captain Marvel and include a riddle by Sir Walter Raleigh and a poem by Fulke Greville, Lord Brooke.

30

HAPPY CHRYSANTHEMUM

The dialogue of interactive fiction is self-shaping. The statement made by the "reader" determines how the computer replies. The technique was first developed for a still-famous dialogue program written in 1966 by Joseph Weizenbaum of MIT. ELIZA was based on nondirective psychotherapy. She never contradicts or challenges. Instead she leads and encourages, often by repeating phrases spoken to her. Thus, if you say "I like bananas" to ELIZA, she is likely to answer "Tell me about bananas" or "Do you really like bananas?"

ELIZA moves us from novel-like works toward pure computer generation of language. A brilliant variation on ELIZA is RACTER (short for "raconteur") created by William Chamberlain and Thomas Etter. ELIZA is designed to elicit comment from the user; RACTER, conversely, is designed to dominate conversation. He has a repertoire of conversational gambits and can draw on a considerable body of topically linked data—a kind of zany hypertext—to enlarge his comments, and he likes to personalize dialogue by continuously using scrambled fragments of comments made by the person with whom he is conversing. The effect is manic rather than Rogerian. Chamberlain collected some of RACTER's happier inventions in *The Policeman's Beard Is Half Constructed: Computer Prose and Poetry* (1984), from which comes the following bit of dialogue. In it one can detect the links between the general topic (music) and the base of musical information through which RACTER crazily gambols, as well as the moments when he leaps wildly from one subject to the next:

[Chamberlain summons up RACTER on the computer screen.]

RACTER: Have we spoken before?
C: No.
R: What is your sex?
C: Male.
R: What is your full name?
C: Bill Chamberlain.
R: What are you?
C: A time-domain reflectometrist.
R: Where are you from?
C: From New York City.
R: I see. You're a time-domain reflectometrist, right? Listen. Let's determinedly cry together. I like captivating Italian music. What kind of music do you like?
C: Jazz.
R: Jazz is O.K. How about classical?
C: I like classical.
R: What about nationality?
C: I also like Italian music.
R: I myself am inflamed by Palestrina. Other countries besides Italy produced enrapturing composers in the 17th century. Sewall was an enraged, important Renaissance master. America was his nation. I take loving pleasure in music. What composer do you like?
C: Telemann.

R: Ah, Bach knew Telemann. A German time-domain reflectometrist of the 18th century liked lettuce. Bach once ate lettuce with Telemann, and this baroque time-domain reflectometrist also desired lettuce. Lettuce sipped with seltzer. Bach liked wine. Do you?
C: Yes.
R: Me too! Telemann loved wine. He loved champagne as well. Telemann once yodeled to Bach, ""Flounder is critical with wine.'' Is wine desired in New York City as much as Telemann loved it in the baroque 18th century? . . .

In general, the more urgently meaning is sought in computer-generated text, the more disappointing the product. Conversely, the less the need to make sense, the more successful the product tends to be, as witness RACTER. In the 1960s, one of the famous literary productions of computers was the sentence "What does she put four whistles beside heated rugs for?" Much more interesting results are produced by a style-analyzing program called TRAVESTY invented by Hugh Kenner and Joseph O'Rourke of Johns Hopkins University in 1984. The program works by analyzing the frequency of letter groups in a sample passage and producing nonsense in the style of the passage. Given a passage from James Joyce's *Ulysses*, TRAVESTY wrote the following:

> Gleaming harnesses, petticoats on slim ass rain. Had to adore. Gleaming silks, spicy fruits, pettled. Perfume all. Had to go back. Had to back. Had to back. His braces all him ass to adore. Gleaming hairness rain yielded. A warm silver, rays of the mutely carved dorn on slim ass rays of the woman plumpness. Uselesh obscurely, he mutely craved down on him assailed. A warm hungered down on him assailed to adore. Gleaming silks, silver, rich, from Jaffa. A warm hungered down on slim braces all. Hig

You might say that TRAVESTY used *Ulysses* to predict *Finnegans Wake*. Engaged in classic activity—stylistic imitation—it also managed to be prophetic.

Computers can write stories as well as sentences. The theory of narratology that descends from the Russian critic Vladimir Propp treats fiction as a sequence of standard plot units. Once identified, the units can be combined and recombined in various, often surprising ways. A computer program using Propp's formulas was created by Sheldon Klein of the University of Wisconsin and a group of associates in the early 1970s. The program can produce hundreds of plotlike scenarios that are combinations of Propp's basic motifs, and it thus serves as an instrument to investigate the form (or forms) of the folk tale. Unfortunately, the variations resemble telegraphic lists of motifs rather than stories in the normal sense, and they soon become tedious to anyone in search of literary effects rather than an improved understanding of narratology. This is not a criticism of Klein, who was interested in how narrative units combine to produce plots rather than in competing with the brothers Grimm. Several later programs have aimed at producing explicitly literary narratives, and some of the results are moderately convincing, although, in general, the longer the story, the more obviously artificial it becomes. Evidently, computers have a long way to go before they threaten the market for the novels of John Steinbeck or Tom Clancy or even for Harlequin Romances.

Poetry has been notorious for obscurity ever since the Oracle of Delphi. Its mission has always been to tell the truth, and this is why it so seldom appears to make sense. Socrates complains in the *Ion* that poets behave like drunkards and madmen, and Dante believed, along with most Christians of his age, that the truths of poetry are veiled in three levels of allegory.

Whether or not obscurity is approved in poetry, it has always been tolerated, and that is probably why computer poetry is easier to accept than computer prose, although the limits of tolerance remain narrow. Permutation is a typical device of computer poets. It is illustrated by Edwin Morgan's justly famous "Computer's First Christmas Card":

The Computer's First Christmas Card

jollymerry
hollyberry
jollyberry
merryholly
happyjolly
jollyjelly
jellybelly

bellymerry
hollyheppy
jollyMolly
marryJerry
marryHarry
happyBarry
happyJarry
boppyheppy
berryjorry
jorryjolly
moppyjelly
Mollymerry
Jerryjolly
bellyboppy
jorryhoppy
hollymoppy
Barrymerry
Jarryhappy
happyboppy
boppyjolly
jollymerry
merrymerry
merrymerry
merryChris
ammerryasa
Chrismerry
asMerryChr
Ysanthemum

In *Cool Pop* Marc Adrien of the Vienna Institute of Advanced Study combines graphics with words in a manner that recalls concrete poetry. Here the visual element is important. A well-tempered computer could produce a large number of such poems, each of a different flavor, and each could be signed by the programmer and adorn the reception room of a great corporation. *Cool Pop* is related by its reminiscence of concrete poetry to the inspired calligraphic art of Scott Kim as in *Infinitely Spiraling Infinity* (p. 212) and to the more obviously concrete style of Aaron Marcus, as in *The City Sleeps but Someone Is Watching* (1972, p. 273).

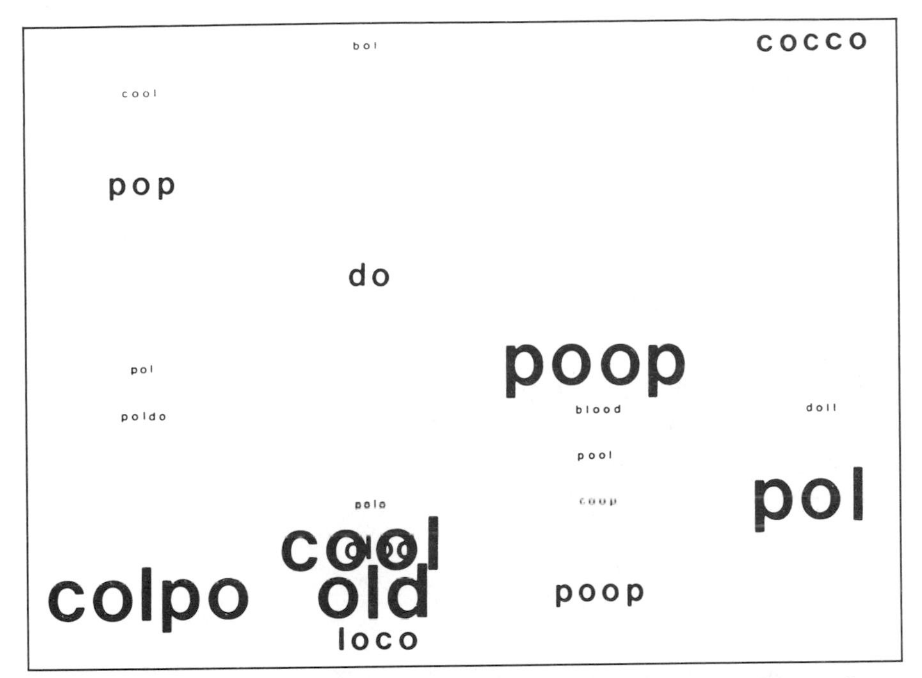

Marc Adrien, Cool Pop *(1966). Note the affinity between this composition and concrete poetry.*

Aaron Marcus, The City Sleeps but Someone Is Watching *(1972). Computer poetry that combines calligraphy and architectural imagery.*

Sustained passages of poetry can be produced by computers. They tend to lean heavily on the technique of repeated combinations of elements evident in "The Computer's First Christmas Card." They also tend to rely on standard grammar and syntax, to use preselected groups of words and phrases, and to follow loosely the rules of verse forms that do not require precise metering or rhyme. They are classic rather than expressive, and they imitate poetry in the way that the early Moog synthesizer imitated an organ. What could be more traditional, more classic, than a haiku? Writing haiku is an old computer trick. Margaret Masterman and Robin Wood of the Cambridge Language Research Unit produced dozens of haiku-like poems in the very dawn of the computer age, of which the following is representative:

all green in the leaves
I smell dark pools in the trees
crash the moon has fled

John Morris of the University of Michigan produced more interesting haiku using a more relaxed program:

Still midnight, silent
Still waters, still frozen,
Battle, dusk, and fear.

These poems have charms, but you might do better writing haiku in the traditional way. Louis T. Milic took a different tack and had better results with a program called RETURNER, which uses a poem by Alberta T. Turner for input. The first few of the several hundred stanzas of output are:

In the morning crowbars will be nearly round,
Separate blankets never step again.
Tomorrow I will ring him through the willows.

Do mice sometimes become like deer at home?
Hemlocks hiss from salad to salad now
But yesterday he often pawed all the apples at the milk pan.

In the morning the quicksand will seem silver
A thin curd also staggers again.

At home the weathercock never appears like the boy
Yet does my bowl turn from salad to salad now?
Carefully we sometimes paw the front porch at the milk pan.

In the morning locusts will turn nearly gray,
Apple twigs often crack again,
So last night they took them through the willows. . . .

Commenting on RETURNER's verses, Turner observes that she had planned a sequence of calendric stanzas in which the year's cycle would be played out in a "threatening tone." The computer poem is more upbeat, even though its "playfulness" is offset by suggestions of "horror." Turner is intrigued by an obscure statement in one of the later stanzas turned out by the program: "Yesterday he sometimes shelled my sisters at the milk pan." Suppose the line had appeared in a poem by Dylan Thomas, who was capable of a maddening degree of obscurity. An interpretation would have been mandatory. Is it not proper to interpret RETURNER? For example, could "shelled" mean "fired a rifle at"? Or might it refer to a glance so intense that it seems to undress—hence "shell"—the sisters being scrutinized? Has RETURNER written a description of the life of a hardscrabble family in west Texas in the 1880s? Or has it written a comment on female awakening? Turner prefers the second interpretation: "The mice-deer images and the shell-at-milk pan images become metaphor for the male invasion and freeing of the young, about-to-mature female personality."

Turner asks, "Just who is the author of RETURNER? I am initially responsible (consciously) for its concrete details. . . . Dr. Milic is responsible (consciously) for retaining these qualities and adding surrealism. The 360/50 computer is responsible (unconsciously) for the sequence and combination in which these qualities are manifested. . . . Each reader is responsible (mostly unconsciously) for the ways in which each part of the sequence will reorder the materials already in his experience bank. RETURNER is therefore poetry insofar as the reader finds that it reorders his own experience to surprise, delight, illumine." We are confronted again with the image of the disappearance of the author and with a text that digests its reader.

Inspired by RETURNER, Turner was able to recast lines that had previously seemed used up. "Instead of taking over from the poet," Turner writes, "the computer freed her to make the poem even more of itself than it would otherwise have been." Thus out of RETURNER she created "Hoeing Song":

At home the rabbits hiss a salad,
chisels chip at a thick curd,
My sisters stick, my bowl staggers,
willows shell me to the pond.

At the milk pan deer are crowbars,
at the milk pan melons sob.
My child has seen me through the willows,
yet I never planted him. . . .

My son has hissed me from the salad,
Apple twigs squirt porcupines.
Rabbits at the milk pan, Daughter,
rabbits shake the blankets down.

A program called MUSESTORM comes closer to transparency than RETURNER because, aside from having a limited pool of words to draw on, its only constraint is that each line must be scannable iambic pentameter. The program supplies its own titles.

Simple B

That simple thou I on for rose in thou,
I seeing thou that never made its sky,
The sky them on if selfsame why them there
That power seeing own then own I own,
And knew that thought the never simple sky.

The words strain to achieve the condition of revelation, and who is to say that they do not succeed?

A computer poem does not have to be considered sacrosanct. It can be "found" among many unexpressive lines, and it can be revised by the poet in the manner of "Hoeing Song." The resulting production is a composite of human and machine input—a literary cyborg. "Red Barrel" was "found" in a composition of seventy lines and slightly modified by a human collaborator:

Red Barrel

I barrel through the red absurd and weep
Although I am unable to abate.

Computer poetry tends to be expressive, but occasionally, it has classic elements. What could be more classic than a sequence of love poems? Louis Milic created the sequence entitled *Erato* by accepting the decree of the computer but added his own insights to it. For example:

Awake

Above the hungry garments of the shore—
Darling, however my spirit can laugh,
Darling, if my life can jump—
Above the humid ruffles of the tide,
Come to me in the pause of the evening.

I have believed some hysterical girls protest,
I, a sure father in the office of the breed,
Still here awake beside the moon.

What could be more somber than the thought of atomic annihilation? In 1962 Nanni Balestrini created one of the most powerful and truly moving of computer poems. It is a meditation on the first atomic bomb, created with a program that uses short passages from Michito Hachiya's *Hiroshima Diary*, Paul Goldwin's *The Mystery of the Elevator*, and Lao-tzu's *Tao-te ching*. The program combines random phrase selection with rules forbidding certain phrase combinations and repetition of phrases within a stanza. Two stanzas are quoted as translated by Edwin Morgan:

Hair between lips, they all return
to their roots, in the blinding fireball
I envisage their return, until he moves his fingers
slowly, and although things flourish
takes on the well known mushroom shape endeavoring
to grasp while the multitude of things
comes into being.

In the blinding fireball I envisage
their return when it reaches the stratosphere while the multitude
of things comes into being, head pressed
on shoulder, thirty times brighter than the sun
they all return to their roots, hair
between lips takes on the well known mushroom shape.

Balestrini's poem is the product of deep collaboration between human and machine. The program that made it began with phrases produced by humans—the lines from *Hiroshima* and *The Mystery of the Elevator* and the *Tao-te ching*. At the end of the output, the stanzas were reviewed by the poet, and the unexpressive ones were culled. Without the original stock of phrases and the human agent, the results would have been trivial. Perhaps successful computer literature requires collaboration. Is the need greater in the case of literary art than in the case of drawing and painting? Or are people simply more conservative about granting freedom to language than about granting it to visual images?

One of the most serious computer poets of the 1980s is William Dickey, winner of a Yale Younger Poets prize. Dickey is fully aware of the precedents and implications of the sort of poetry he is composing. He is concerned with space on the page as well as with words, and also, apparently, with sound. "She Rescued Him" owes an indirect debt for its sense of the possibilities of spatial arrangement to Mallarmé's "Un Coup de dés." Its own victory over silence is announced in the last line:

She Rescued Him From Danger as the Bell

clanged.   Midnight.
always in these stories.
        Real Life   organizes
                for dawn &
        not
                so simple.
                    Which
                will wake
first   make
    orange juice   which
            will say

            the spell the
                awakening words?

31

AT THE TOP OF THE MASTHEAD

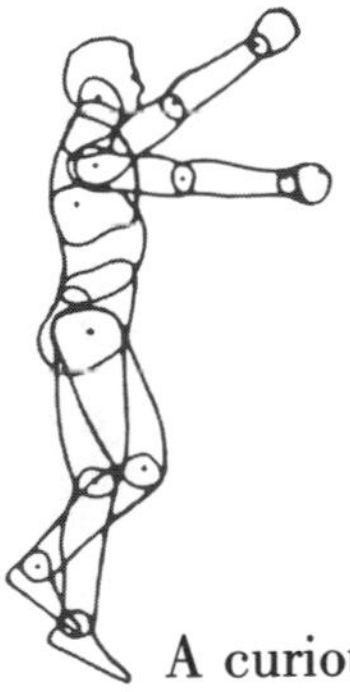

A curious feature of all computer art is its ambivalence regarding human input. Should the human artist retire gradually to allow the computer to develop its intrinsic aesthetic capability as Robert Mallary advises in connection with his six stages of computer sculpture? Should the artist assert the human presence vigorously, as in the case of interactive fiction? Should the artist seek to vanish entirely, as happens in Robert Parslow's and Michael Pitteway's "Random Drawing"? Or is there an ideal symbiosis between man and machine?

Interactive man-machine collaboration seems more urgent in literary art than in other art forms. Louis Milic's RETURNER and Alberta Turner's "Hoeing Song" and Nanni Balestrini's "Hair Between Lips" show what a broad range of effects this sort of collaboration can achieve. Pierre

Boulez mixes human voices with computer-generated instruments, and Harold Cohen uses a Logo turtle to create the outlines of his murals but does the coloring by hand, while other artists use computer techniques to deform photographs. Perhaps collaborative art of this sort expresses humanity's confidence in itself—confidence that it interacts with its creation as an equal and that human criteria continue to be essential to art. Perhaps, as Robert Mallary insists, collaboration is a transient phenomenon, a by-product of fourth-phase symbiosis between man and machine. In the fifth of Mallary's phases the only action the artist can still perform is to pull the plug, and why should the artist want to do that?

Machines have always seemed to imitate humans. They are, in one definition, merely extensions of human functions. Science fiction objectifies subconscious anxiety about machines by showing people victimized by robots and by androids indistinguishable from humans. In most such fiction, because of attributes considered impossible for machines, the humans win at the last minute. But the relationship of man and machine is more intimate than the metaphor of robot war suggests.

In one of Rod Serling's "Twilight Zone" episodes, a man is injured in an automobile accident. His arm is badly cut. He opens the folds of the cut. Inside his arm are transistors and wires. He is stunned. He had always assumed he was human. Now he must acknowledge that he is an android. Should he tell his family? Are there any human beings left? At some time in the future that question may become unanswerable, but by that time it will not make any difference.

Herman Melville describes the tendency of the lookout on a Pacific whaling boat to drift into a trance in which he takes "the mystic ocean at his feet for the visible image of that deep, blue, bottomless soul . . . and every strange, half-seen, gliding, beautiful thing that eludes him; every dimly-discovered rising fin of some indiscernible form, seems to him the embodiment of those elusive thoughts that only people the soul by continually flitting through it."

Is the traditional concept of art still adequate for the images and music and poetry they are creating? What is the meaning in computer music of terms like "compose" and "instrument" and "perform"; in art of "artist" and "paint" and "model"; in literature of "author" and "write" and "read"? When computer art is interactive, does the consumer become the artist?

When computer art is collaborative, is the machine equal in standing with the artist? Is artificial reality a work of art or is the individual who is enveloped in it the work of art?

Perched on the topmost spar of technology, computer artists do not need to answer. They gaze out into hazy, blue, fathomless realms of abstraction, fascinated by strange, half-seen, gliding, beautiful shapes they have created.

PART V

DEUS EX MACHINA; OR, THE DISAPPEARANCE OF MAN

Consume my heart away, sick with desire
And fastened to a dying animal.

—W. B. YEATS

Peter Paul Rubens, Ganymede *(1602). Man lifted to the gods.*

32

THE CURVE OF EVOLUTION

The curve of evolution begins to rise slowly, almost imperceptibly, but its angle of ascent constantly increases, and the rate of increase is exponential. It took around three billion years—perhaps two-thirds of the total age of the earth—for the molten planet to cool off and the seas to form and for the earliest single-celled organisms to appear. About 600 million years ago, they were joined in the warm, shallow seas of the Cambrian Period by multicelled creatures that excreted external skeletons. Within another 250 million years, animal life had crept onto the beaches and into the forests of giant fern and ginkgo trees. Pangaea appeared, the land mass from which today's continents were formed.

By the Triassic Period, a scant 250 million years before the beginning of the present era, the first mammals and dinosaurs had appeared. No more than 100 million more years were needed for the appearance of birds and small warm-blooded animals, and as the dinosaurs left the evolutionary stage with an abruptness that is still not fully understood, plants began producing flowers to brighten the dark days. Seventy-five million years later, in the Oligocene Age, whales began to clear their spouts on the oceans, while on land anthropoid apes swung from deciduous trees and huge herds of grazing animals roamed the broad savannas. Within another twenty million years the apes and man had parted company. An evolutionary phase transition had occurred.

With this event interest shifts from the curve of evolution in general to the equally dramatic curve of human evolution.

Thanks especially to the discoveries of Louis and Mary Leakey, a good deal is known about what Charles Darwin called the descent of man. In fact, discoveries about the earliest appearances and diffusion of hominids are occurring so frequently that any chronology—including this one—risks being obsolete by the time it is published. According to present evidence, *Ramapithecus*, the earliest hominid, appeared in the late Miocene period around 12 million B.C. *Ramapithecus* lived on the ground rather than in trees and was probably semi-erect and vegetarian. The species ranged from Europe to Africa and east to India. *Homo erectus* appeared ten or so million years later—definitely a more enterprising sort of creature who used fire and, in the opinion of some, practiced ritual cannibalism. Perhaps *Homo erectus* had developed the rudiments of a mythology in which terrible or benign figures with human qualities stalked the earth. His traces are found all the way from Africa and Spain to eastern China and Indonesia.

Roughly a million years separate *Homo erectus* from modern man—*Homo sapiens*—whose earliest unambiguous performance on the evolutionary stage is breathtakingly recent—around 100,000 B.C., at which time he hunted woolly mammals and competed with Neanderthal Man for the honor of becoming the sole possessor of reflective intelligence on earth. *Homo sapiens* was a hunter and gatherer. Although his tools were initially unpolished stone, he soon learned to transform the unpolished into the polished and to make brilliant ritual paintings of the animals he hunted. He certainly knew the rudiments of a religion and almost certainly imagined creatures like himself but different who were alternately his helpers and his tormentors. He learned to worship them and to use

African mask from the National Museum of African Art, Washington, D.C. (undated).

incantations and rituals to control them, and he often represented them in masks and in performances of ritual dramas.

An evolutionary leap—a new singularity—is associated with *Homo sapiens*. While man has remained much the same physically for the last 40,000 years, human culture has evolved along an exponential curve like the one describing biological evolution. The move from hunting and gathering to agriculture began around 10,000 B.C. and took most of human history. The earliest cities date from around 6000 B.C. Early Egyptian and Cretan civilizations began between 4000 and 3500 B.C., during which period copper gave way to bronze. The so-called Legendary Rulers of China governed from 2850 to 2200 B.C. And writing appeared: Sumerian and Egyptian hieroglyphics date from around 3600 B.C.; the Phoenician alphabet, the ancestor of the modern phonetic alphabet, was being used by 1100 B.C.; and the Indian Vedas date from 1000 B.C. Paper first appeared in China in 950 B.C.

Some three thousand years separate the early Near Eastern and Egyptian empires from the Roman Empire. About a thousand years separate the conversion of Constantine from the Renaissance. Columbus died some 250 years before the beginning of the Industrial Revolution. Moving from the Industrial Revolution to the age of steam took a century, while the move from steam to electricity and the internal combustion engine took perhaps fifty years. The next fifty years saw the arrival of radio, television, jet aircraft, and atomic fission, followed within two decades by space flight and genetic engineering.

Cultural evolution should not be understood, any more than biological evolution, in terms of movement from bad to good or good to better. Its absolute direction is best symbolized by an arrow pointing down a dark corridor.

There is, however, a unifying theme. Every advance in culture has been an advance in communications and has encouraged ever-larger organizations of the human beings who produced it. Small, isolated bands of hunters combine into tribes which form city-states which in turn form—or are absorbed into—nation states and empires. The movement has downs as well as ups, but the direction is continuous. Twentieth-century technology has perfected the work of Renaissance explorers, Enlightenment scientists, and Victorian entrepreneurs. Teilhard de Chardin writes in *The Future of Man:* "No one can deny that a network (a world network) of economic and psychic affiliations is being woven at ever-increasing speed which envelops and constantly penetrates more deeply within each of us. With every day that passes it becomes a little more impossible for us to act or think otherwise than collectively."

Teilhard, who sees the process as a kind of continuous Revelation, argues that with the advent of mind there is "outside and above the biosphere . . . an added planetary layer, an envelope of thinking substance, to which . . . I have given the name of the noosphere." Eventually, he proposes, this "envelope" will become a seamless web of relationships uniting all men in global communion. Its collective expression will be a human song responding to the inaudible music of the voice of God. The unity of this condition will be much like the unity understood in the Mass as incorporation into the Body of Christ: "The idea is that of the earth not only becoming covered by myriads of grains of thought, but becoming enclosed in a single thinking envelope so as to form, functionally, no more than a single vast grain of thought on the sidereal scale, the plurality of individual reflections grouping themselves together and reinforcing one another in the act of a single unanimous reflection."

For many this is airy nonsense; for others it is an inspiring vision. It is, at any rate, a very clear statement of the idea that man—at least man in the old sense of a separate and individual essence—may disappear as a result of evolution. The notion of human transcendence is ancient. It is foreshadowed by the ancient myth of Ganymede, who was captured by Jove and brought to heaven to be his cupbearer.

33

FROM AJAX TO ENIAC

Meanwhile, another kind of evolution has been occurring. I am speaking of the evolution of silicon intelligence—or, as the earliest popular metaphors had it, "electronic brains."

"Electronic brain" is a metaphor of life, and let us be clear about this: it *is* a metaphor. It does not mean that computers are alive, any more than "My love is like a red, red rose" means that my love has a scarlet face and is covered with brambles. Metaphors and similes say what things are "like" and suggest emotional responses. They are useful for explaining and clarifying because they can say how something that is not understood resembles something that is understood. They are not, however, to be taken literally.

Human beings have always been fascinated by gadgets and toys that

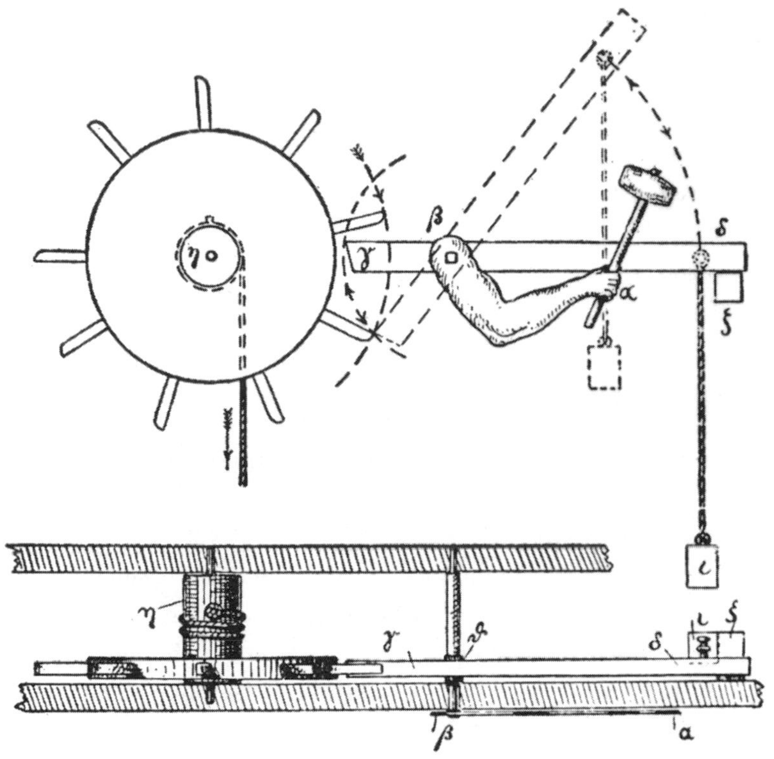

Mechanical arm by Hero of Alexandria (4th c. A.D.*) as drawn by Robert Brumbaugh (1966).*

are "like" living creatures. Perhaps the toys are benign images of the alternately cruel and helpful supernational beings with which mythology peoples the earth.

Simple mechanical devices created from cogs, levers, springs, and screws are capable of endless surprising variations on their simple norms. Who has not been enthralled as a child by a windup cow that moos or a dog that wags its tail? By a doll that cries and bobs its head and drinks milk and wets its pants? By a toy soldier that marches across the table firing a gun?

In the fourth century A.D., Hero of Alexandria catered to just this human delight in toys that seem to be alive. Consider his aviary as described by Robert Brumbaugh in *Ancient Greek Gadgets and Machines.* When water enters its lower chamber, air is expelled through the throats of the birds. They warble beautifully unless an owl is turned toward them. If the owl *is* turned toward them, they fall silent. William Butler Yeats is recalling an aviary like this when he asks in "Sailing to Byzantium" to be like a bird "set upon on a golden bough to sing/To lords and ladies of Byzantium."

Hero's mechanical theatre is his most ambitious experiment in the imitation of nature and life. In fact, its complexity anticipates the idea of "artificial reality" in computer art. When a trumpet is sounded, the doors of the theatre open and mannequins perform a five-act tragedy. The play opens with nymphs building a ship using hammers, saws, and drills. When the ship is launched, dolphins can be seen leaping from the sea. Later there is a storm with thunder and lightning. The ship is wrecked and by the end the protagonist has drowned. Hero is fascinated by the "likeness" to life of this play. Describing the launching of the ship, he pauses to exclaim that the dolphins appear and disappear in the sea "Just as they really do!"

The Greeks were already aware that the metaphor of machine life cuts two ways. If machines can be like people, people can be like machines. A fourth-century B.C. author quoted by Brumbaugh remarks, "People respond to the impressions that strike their senses like self-moving puppets."

The history of things that are "like" humans has its mythic as well as its mechanical side and its dark as well as its playful aspects. There are innumerable legends of magicians who create slaves from inanimate materials and heads of brass that can foretell the future, and in most of the legends the creations begin well but end in tragedy. The point is illustrated by the most famous of all such legends, the legend of the Golem. *Golem* means "an unformed thing" in Hebrew. It is used in Genesis for the handful of dust into which God breathes His spirit to create Adam. According to a common version of the legend, the Golem was made by Rabbi Loew of Prague, who was a friend of the astronomer Johannes Kepler, to protect the Jews of Prague from pogroms. It had superhuman strength. Soon it turned terribly destructive and had to be destroyed. Mary Shelley was thinking of the Golem legend when she invented the plot of *Frankenstein*. The story turns up again and in Prague in the twentieth century as one of the sources of the play in which the word "robot" was first used—Karel Čapek's *R.U.R.* (1921).

"Cybernetics" is a term introduced by the mathematician Norbert Wiener to define a new discipline concerned with self-regulation in machines.

Hironymus Bosch, Devils Tormenting Men. *Detail from* The Garden of Delights *(ca. 1500). Man has always felt he shares the earth with invisible beings, some benign and some intent on tormenting and destroying him.*

The word is from *kubernetes,* which is Greek for "helmsman," and refers to feedback, which is the means whereby a machine regulates itself. The word *governor,* used to mean a device that permits engines to be self-regulating, derives from *kubernetes.*

Cybernetics grew out of Wiener's interest in the "likeness" between animals (including human animals) and machines. The subtitle of *Cybernetics* (1948) is *Control and Communication in the Animal and the Machine.* Wiener stresses two kinds of "likeness." First, the central nervous system employs "circular processes, emerging from the nervous system into the muscles and re-entering the nervous system through the sense organs." These are "like" the feedback of a cybernetic machine. Second, for Wiener the firing of an organic neuron is "like" the "on-off" switches used in electrical circuits and digital computers: "The all-or-none character of the discharge of neurons is precisely analogous to the single choice made in determining the digit on a binary scale . . . and must have its precise analogue in the computing machine."

There is something about cybernetic machines that invites comparison to living beings. Are such machines in some way aware of themselves? Obviously not, but they behave in ways that mimic the behavior of organisms that have self-awareness. They are, you might say, a three-dimensional metaphor for self-awareness.

The escapement mechanism that allows mechanical clocks to tell accurate time appeared in Europe in the thirteenth century. Its function was to regulate the motion caused by descending weights and later (by the fifteenth century) coiled springs. Thanks to its escapement, a clock behaves differently from the simplest machines.

A clock with an escapement mechanism might be compared to a protein suspended in a Precambrian ocean. It is rich with potential. Within a century after the first escapement, clockwork mechanisms—in addition to telling time—were being used to create automated animals and people who acted little plays much like those of Hero of Alexandria for delighted townspeople who turned out to watch their daily performances on church steeples and *Rathaus* towers. The metaphor of the likeness of clocks to living things was thus objectified by clockwork figures that had no use other than to enact the metaphor. As technology improved, the likeness improved along with it. By the mid-eighteenth century Jacques de Vaucanson had created an automated flute player who covered the stops on a flute with mechanical fingers and was capable of playing twelve different tunes.

Precise measurement of time is essential to exact observation of real-world events and also for navigation, since determination of longitude depends on time measurement. Since precision is essential in clock-making, the increasing demands placed on clocks accelerated the development of machine tools. And since a clock is an elementary computer, development of clocks is related to the development of schemes for mechanical computation.

By permitting precise control of time, clocks also changed the nature of work, turning the day into standard and repeatable segments and permitting wages to be related to hours of work. Clocks therefore contributed mightily to making human behavior more "like" the behavior of machines. This is a significant point because the introduction of machines that are like people into society is usually considered a one-way street. It is not. Recall the comment, quoted by Robert Brumbaugh, comparing people to puppets. The more that humanlike machines become a part of human culture, the more human culture changes in response to their presence.

Another step in the development of machines that are like people was the calculating machine. In the seventeenth century Blaise Pascal invented the first practical device of this sort: the Pascaline, which could add and subtract handily and with effort could also do basic multiplication and division. Half a century later Gottfried von Leibnitz, coinventor, with Sir Isaac Newton, of infinitesimal calculus, solved the problem of multiplication and division. New dimensions were added to the project at the beginning of the nineteenth century with the invention by Joseph Marie Jacquard of a loom that employed a system of punched cards to automate weaving patterns.

Charles Babbage and his collaborator Ada—Countess of Lovelace and daughter of the poet Byron—combined many technologies that had been developing separately during the preceding century to create the first realistic design for a general-purpose mechanical computer. The initial object was to produce accurate tide tables, and the engine was planned with a printer to output the tables.

Like some wonderful Jules Verne fantasy, it was to be powered by steam. Among its components were a "mill" which did calculations and a "store" for memory. Input was by punched cards like those used by Jacquard. Ada wrote in her *Observations on Mr. Babbage's Analytical Engine*, "The Analytical Engine weaves algebraic patterns just as the Jacquard loom weaves flowers and trees." The comment is intriguing because it unconsciously anticipates the idea that conceptually a computer is a universal machine. A sophisticated computer is as capable of making images of flowers and determining weaving patterns as solving mathematical equations.

If he had succeeded, Babbage would have achieved something foreshadowed for centuries in mythology, toys, gadgets, and machine design. The analytical engine was intended to think. It would think exclusively about numbers, but it would think about them intelligently enough to say things that were vitally important to its human attendants. Unfortunately, it was a failure. Mechanical devices can add, subtract, multiply, divide, and sort within limits, but the limits are narrow. When they become complicated, they encounter the liabilities of all mechanical systems: inertia, friction, inaccurate machining, slippage, distortion, fatigue, breakage. Babbage's prototypes were given to fits of shuddering and wrenching as the effects of mechanical imperfections compounded themselves during operation.

Ada died at thirty-six of cancer in a room padded with mattresses to protect her during violent spasms of pain. Babbage lived on, a failed visionary. He was, nevertheless, regarded with awe. When he died at eighty his brain was examined by the eminent surgeon Sir Victor Horsley, and it can still be seen in two jars, one per hemisphere, in the Hunterian Museum of the Royal College of Physicians.

A century later, under the pressure of such urgent tasks as deciphering enemy codes, tracking high-altitude aircraft with gun batteries, and creating an atomic bomb, a series of dramatic advances occurred. Among those responsible for them are some of the most revered names in twentieth-century science: Alan Turing, John Von Neumann, Norbert Wiener, Claude Shannon, and Herbert Simon, to name only a few. These men understood that electrical currents rather than physical motions of cogs and levers and wheels would be the future carriers of machine intelligence. Wiener was fascinated by the likeness between the carbon-based circuits of the human neural network and wired circuits. Although the analogy did not lead immediately to useful applications, the fact that neuron circuitry existed and obviously worked bolstered faith in the possibility of using electronic circuits to create intelligent machines.

The immediate ancestry of modern digital computers is well established in Germany and England. The first Germans to build a prototype of a modern computer were Konrad Zuse and Helmut Schreyer. Fortunately for Western democracy, Zuse and Schreyer were unable to interest the Nazis in supporting their research. In England, although theoretical work had been done before the war, it took the urgency of the life-and-death struggle with Germany to produce practical results. At the beginning of the war, the British had obtained a German code machine called Enigma from the Polish Secret Service. At an English country house called Bletchley Park, a series of machines was secretly created to decode German signals. The final product of the design effort was a digital computer called COLOSSUS, which became operational in 1943. COLOSSUS was extraordinarily fast because it used vacuum tubes rather than relays. According to many intelligence experts, COLOSSUS may have provided the advantage that shifted the course of the war in favor of the Allies.

In America the first practical electronic computer was the work of John V. Atanasoff, a professor of electrical engineering at Iowa State University, and his assistant Clifford E. Berry. Atanasoff did not invent the key concepts needed for digital computing. These had already been developed by George Stibitz of Bell Telephone Laboratories and Howard Aiken of Harvard, who went on to produce the Mark I computer, a cumbersome special-purpose machine using electrical relays. Atanasoff's achievement was the creation of a general-purpose computer that, like COLOSSUS, used vacuum tubes. Although a first model, called ABC, was completed in 1940, Atanasoff abandoned his work on computers during the war, and credit for developing the first practical digital computer was, until recently, assigned to John W. Mauchly, creator, with J. P. Eckert, of ENIAC, which became operational in 1946. Later, when John Von Neumann suggested that the computer could store instructions as well as data, the basic elements of the modern computer were in place. Equipped with memories that could store instructions, computers took a giant step toward being self-reflexive. They began to assume control of their own operations.

In spite of the apparent complexity of their spaghetti-like circuitry and glowing vacuum tubes, the first electronic computers were more like the earliest protozoans than advanced organisms. By 1960 they had progressed from vacuum tubes to transistors to silicon chips. They had slimmed down, and although their circuitry constantly grew more complex, they lost their clumsy appearance. They looked increasingly elegant as they floated lazily on the ocean of possibility out of which they had come. They were not very smart, but they were smarter than any of the nonhuman devices that preceded them. And they continued to evolve. As their dimensions decreased, their capacity grew by orders of magnitude.

In *The Micro Millennium* (1979), Christopher Evans provides a colorful measure of the progress between 1950 and 1978. He assumes that a human brain has roughly ten billion neurons. This is a low figure since

Minimal robot—rhino robot arm (1987).

neurologists put the number closer to fifty billion, quite aside from the mode in which the connections are made, which makes for something like 10^{16} connections, but let it stand. A computer having ten billion circuits and using the vacuum tubes of 1950 would be as big as New York City and would consume as much power as the entire New York subway system. It would also be a piece of junk because the tubes would continually be burning out. With the coming of transistors in 1955 the space requirements for Evans's machine shrank to roughly the size of the Statue of Liberty. With the large-scale integrated circuits of the early 1970s, Evans estimates that his machine could be packed into a space the size of a bus. With very large scale integrated circuits, his machine is the size of a television set and uses about the same amount of power.

You do not have to buy Evans's dimensions to understand his main point. The reduction in scale of computers since 1950 has been exponential. At the same time, the increase in their abilities measured both in millions of instructions processed per second (MIPS) and in memory size has also been exponential. Costs have also fallen dramatically. An analogy is sometimes used in the computer community: if the cost of transportation had fallen as dramatically since 1950 as the cost of computing power, today's traveler would be able to buy a round-trip ticket to Mars for $12.50.

Something else happened as miniaturization progressed. Computers ceased to resemble single-cell organisms. They began to resemble small, multicelled colonies, an analogy made more apt by the use of co-processing.

Carbon life took more than a billion years to progress from single-celled to multicelled creatures. Silicon devices managed something similar in twenty-five years. They were able to move fast because they were, in a sense, spiritual parasites: they drew their understanding predigested from their hosts. Their feeding, healing, and reproductive functions were all supplied for them. It is as though carbon creatures had developed brains and sense organs before they had begun to grow bodies.

34

LEARNING TO TALK

A child is born with the capacity for speech. The world programs it. Is it taught to speak or does it learn to speak or does a structure hard-wired into the brain respond to stimuli like the strings of a harpsichord vibrating at predetermined frequencies when plucked by a quill?

Alan Turing, a brilliant mathematician and a member of the team that produced one of the first working digital computers—COLOSSUS—proved in 1937 that the simplest tape-punch machine can be a powerful computer if it can track long enough and keep accurate count of the holes and blanks in the tape. Of course, no such wonderful machine existed or could be built. The Turing engine is a theoretical model, not a realizable project. However, even a fairly primitive computer has the power to answer urgent questions. In the period between 1940 and 1950, especially urgent questions clustered around code breaking and making a uranium bomb and then a hydrogen bomb, and on airfoils, artillery fire, and rocket trajectories.

The silicon devices of the 1950s taught their programmers how to converse with them in the equivalent of creole dialects—awkward mixtures of machine and human language. Some of the dialogue had to be carried on by physical rewiring of the consoles. Later, the talk was in code that could be communicated to the computer by punched cards and tape. Eventually keyboards, magnetic tape, and disk drives became the standard methods of input.

By the mid-1950s, programs were being created that exhibited limited forms of what looked like human intelligence. A watershed for these developments occurred in 1956 at a conference at Dartmouth. The conference was convened by two younger scientists, Marvin Minsky, then at Harvard, and John McCarthy, then at Dartmouth, and two senior figures, Claude Shannon of the Bell Telephone Laboratories and Nathaniel Rochester of IBM. It lasted for two months. Out of it came an agenda for future computer research and a name: artificial intelligence.

Artificial intelligence—usually called AI—is the investigation of ways in which computers can be coaxed to emulate what would be considered intelligent behavior in humans. Its often stormy history has been written several times, most usefully, perhaps, by Pamela McCorduck in *Machines Who Think* (1979). Its development is closely associated with MIT, Stanford, and Carnegie-Mellon University. Among the key players were (and are) Marvin Minsky, Seymour Papert, and W. Daniel Hillis of MIT; Herbert A. Simon, Allen Newell, and Hans Moravec of Carnegie-Mellon; and Edward Feigenbaum, John McCarthy, and Terry Winograd of Stanford. In addition to being a pioneer in artificial-intelligence applications, McCarthy is the author of Lisp, the computer language in which much artificial-intelligence programming is done. Hillis, a Minsky student, is the founder and president of Thinking Machines, which makes computers using parallel processing.

The practical career of artificial intelligence began at the Dartmouth conference when Allen Newell and Herbert Simon exhibited printouts from a program they had developed called Logic Theorist. Logic Theorist was able to prove theorems in the *Principia Mathematica*, a fundamental work on mathematical theory by Bertrand Russell and Alfred North Whitehead. It should have created a sensation at the conference. For one thing, it showed that programs with artificial intelligence could actually be created. For another, it showed how a program could be given an inner direction—an ability to search out answers on the basis of a general goal. The excitement did not happen. Although Logic Theorist was politely noted, it was less than a sensation, and Simon and Newell were disappointed.

In spite of the letdown, they followed Logic Theorist in 1957 with a more powerful program called General Problem Solver. Pamela McCorduck calls this program "the first . . . ever developed as a detailed simulation of human symbolic behavior; as such it clarified . . . a handful of procedures human beings had been using all along for solving problems." The comment is interesting in two ways. The phrase "human symbolic behavior" suggests how tantalizingly close computers seemed to be to human intelligence even before the 1960s. It also underscores a recurrent theme of computer research: the fact that an advance in understanding computers often seems to permit an equally important advance in understanding the mind. The General Problem Solver was well received. A decade after its introduction it was still being upgraded and used in new ways.

Challenged to compete directly with humans, silicon devices quickly learned to play checkers. Thanks to Arthur Samuel of IBM, they were playing championship-level games by 1960. They could also play rudimentary chess. But, said skeptics, they will never be able to play serious chess. That is a skill only humans can master.

The "chess challenge," as it might be called, is an interesting phenomenon. Many scientists are devoted chess players and consider the game a test of intelligence. Surely a good chess player is intelligent in a human sort of way. The development of chess skills in computers illustrates how computers can force their users to revise basic concepts about what is uniquely human. Chess *does* require intelligence, but up to a certain point—a rather advanced point, in fact—the kind of intelligence needed is one that can be simulated by machines. Computers could play elementary chess in the 1950s. By 1975 Richard Greenblatt had created a program capable of "Class C" chess, which is respectable for a human player, if not inspired. By 1978 rival programs had reached a still more impressive "Class B" level. Confronted by this sort of progress, advocates of the "chess challenge" have to admit that at least some kinds of intelligence are not uniquely human because they can be modeled—if not exactly duplicated—by machine programs.

In the early 1970s Terry Winograd began experimenting with a program that could represent its environment and discuss it intelligently. As was noted in the discussion of language in Part III (p. 203), he called the program SHRDLU from the last six of the twelve most frequently used letters in the English language. The newly created environment was called "Block World." The machine was told that it confronted a set of blocks of different colors and shapes, and it was then ordered to manipulate the blocks—for example, to place a red pyramid on a green cube.

SHRDLU had to understand space and gravity. It had to know enough about shapes to realize that the green cube could not, for example, be put on top of the red pyramid. It also needed to understand English as used by its conversational partner and how to reply in well-formed sentences. Furthermore, it had to keep track of the blocks it was moving so that if, for instance, it was asked to move a cube that was already under a pyramid, it would remember to pick up the pyramid and set it aside before trying to move the cube.

The abilities of SHRDLU were quite impressive, but Winograd eventually ran into limits. To go much beyond Block World required far more computer speed and power than were available. Continuing research has shown that to be significantly more intelligent, SHRDLU would also need new programming techniques and probably new types of computer design.

Another program that apparently exhibited considerable intelligence was the Automated Mathematician, created in the 1970s by Douglas Lenat. This program seemed to operate itself using built-in general rules of procedure ("heuristics"). The Automated Mathematician explored set theory, then proceeded to invent arithmetic, and finally carried out an analysis of prime numbers. Lenat's Eurisko improved on the Automated Mathematician by its ability to change its procedural rules in the light of experience. Eurisko is credited with having originated an innovative design for integrated circuits, and in 1982 it was so successful in the war game Traveler that after its second victory it was barred from further Traveler tournaments.

On the other hand, many of the headiest predictions of the early days of artificial intelligence refused to come true. Machine translation is a good example. Generously supported by the Defense Department, early researchers were confident they would produce excellent translation programs before the end of the decade of the 1960s. Part of the reason for their optimism was that they approached the easy problems first, before the true dimensions of the challenge appeared. As they confronted the harder problems, the date for the perfecting of a general-purpose translation program kept receding into the future, and the complexity of the programs and amount of computing power needed to achieve even limited success kept increasing.

The problem of machine translation was still being attacked in the late 1980s. And in spite of renewed enthusiasm among researchers, the final solution was still proving elusive. As with machine translation of language, so, too, with many problems involving what John McCarthy and others called "common sense." Common sense is a faculty humans seem to develop without effort but computers to develop hardly at all. The more that is learned about this human faculty, the more elusive becomes the goal of endowing a computer with it (although that does not, of course, stop people from trying).

So far, the greatest successes of artificial intelligence have been in the area of what are called "expert systems," which use information and rules of procedure drawn from experts in the relevant area of knowledge. Each system is a model of the knowledge an expert would have or need to recover when solving a specific problem. A skilled physician, for example, brings a general knowledge of symptoms and diseases to bear on a sick patient. By using well-formulated rules of procedure, he is able to eliminate certain diagnoses, leaving others to be tested further. In order to decide which tests to administer, he may have to review current medical literature on the diseases for which he is testing. Ideally, the testing should eventually eliminate all but one possibility, which will be the diagnosis. Having reached this point, the physician draws on other knowledge and rules of procedure to prescribe treatment.

An expert diagnostic system will use the resources and emulate the procedures of skilled diagnosticians. It will include a database of medical information, and it will query the patient and draw inferences by following rules of procedure distilled from painstaking interviews with the appropriate experts. Once created, the system must be thoroughly tested and revised in the light of experience. The payoff is that when properly crafted

Robot Boardroom *(Maxell computer disk advertisement, 1987). Is it really all that easy to pull the plug?*

and used with circumspection, expert systems work. They give advice today on everything from medical diagnosis to probable locations of mineral deposits, legal research, maintenance of complex machines, investment strategies, and navigation. Among commercially successful expert systems are Dendral for chemists, Macsyma for mathematicians, Prospector for geologists, and Mycin and Internist for physicians. According to users, these systems can seem to exhibit almost human understanding. Dendral, to which Edward Feigenbaum contributed, is said to be better than human researchers at the task of analyzing the structure of complex molecules.

Still, expert systems are like computer chess programs: they are very good at specific jobs, but they lack flexibility. Their rules and specialized data have to be taught them by human experts. They cannot learn on their own, and as knowledge grows and changes in a given field, they have to go back to school. If the theory on which their methodology is based changes, they have to be reprogrammed. They are not very smart by biological standards in spite of their impressive specialized abilities. Perhaps their IQ is about on a par with the IQ of a Cambrian mudworm.

35

THE RIGHT CONNECTION

The dominant tradition of computers from the mill and store of Babbage's Analytical Engine to the central processing units and random-access memories of today's mainframes has been serial processing. In serial processing, items are lined up at the computer's central processor like people waiting in a line to buy movie tickets. The processing goes forward item by item. The strength of the method is its strict logical sequencing. The limitation is that running every step through a single unit creates an obvious bottleneck no matter how fast the unit. A second and equally troublesome limitation is that serial processing is hierarchical and must be operator-organized. A lot of the world, including, evidently, the human brain, operates on different principles.

In the 1980s something like a phase transition has been occurring in computer evolution, with the rise to prominence of machines and programming techniques that are parallel rather than serial. Parallel processing is named for the fact that it divides a problem into parts that can be treated simultaneously or in self-arranging sequences. This is more complicated than having four ticket booths at the movie theatre rather than one, but that analogy at least suggests how parallel processing can be faster than serial processing.

Parallel systems are based on the likeness of computer circuitry to the circuitry of the brain. They are often called "neural networks," and the nodes that are linked by the net are often called "neurons." Since the connections among the "neurons" are an essential part of the operation of parallel computers, machines designed from the beginning to use parallelism are called "connection machines." Since parallel processing mimics what is assumed to be the manner of operation of the brain, it is not surprising that as parallelism has gained in importance, there has been a renaissance of what has been called "reverse brain engineering"—that is, analysis of brain functions based on the theory that they are "like" the functions of neural computers. At the same time, while stimulating research on neural circuitry in the brain, the rise of parallelism has given new currency to the metaphor of machine life.

Parallel processing has many virtues. Most obvious is the fact that it permits certain tasks to be done more rapidly than serial processing. However, this may not be the most significant advantage. Parallel processing is flexible. In certain kinds of parallelism, the program operates by creating configurations of circuits that are a reflection of the structure of the problem being analyzed. When the problem changes, the configurations change. Depending on how a given program is set up, processing can also be hierarchical—that is, layered—so that information is refined as it moves from one layer to the next. In addition, processing can use feedback. That is, messages can be sent from a higher layer, which is closer to having generalized understanding, to a lower layer, which is closer to the raw data being received.

A recurrent model for this sort of system is provided by what is known about the architecture of vision. The information received by the nerves in the retina of the eye is processed in several stages (or layers) as it moves from the eye to the cortex of the brain. The processing moves generally from "bottom to top," but there is a countercurrent from top to

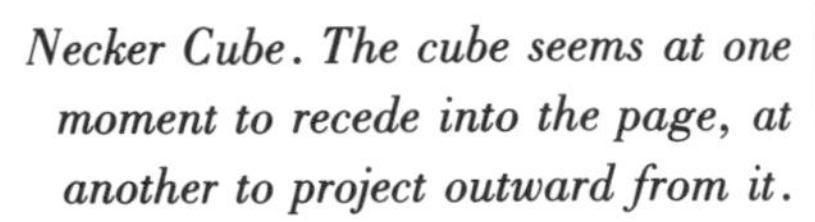
Necker Cube. The cube seems at one moment to recede into the page, at another to project outward from it.

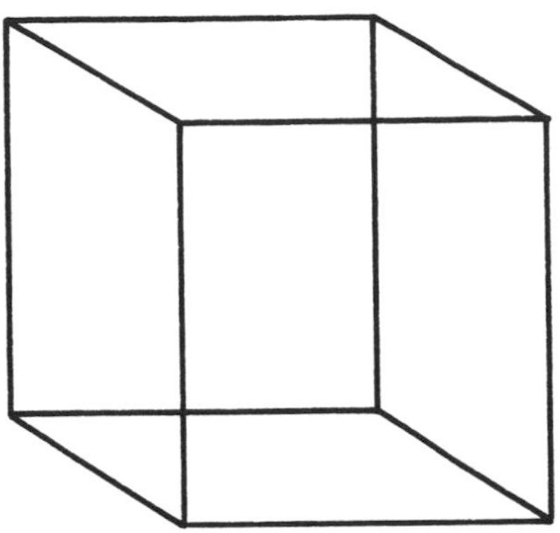

bottom, so that what is "seen" can be influenced by input from the pattern-recognizing part of the brain.

A common example of perception in which immediate input is affected by higher-level processing is the Necker Cube. A Necker Cube is simply a stick-figure cube with all of its lines drawn. The interesting thing about the Necker Cube is that it will not stay still. At times it seems to project out of the page toward you, and then suddenly it seems to pop inward and away from you. Moreover, you do not control which of the two perceptions you have. There is a fifty-fifty choice between the inward-going and the outward-going interpretation of the Necker Cube, and your perception mechanism solves the problem by alternating between versions in spite of any conscious effort you may make to put a halt to the process. Evidently the information being transmitted by the rods and cones of the eye is structured at some higher level, and this structuring influences how and what you "see." Richard Restak observes in *The Brain* (1979), "The perception of reality is best understood as a constructive process by which the brain builds useful models of the world."

"Top-down" input in parallel processing is "like" what apparently goes on in the brain in processing the image of the Necker Cube.

Research on parallelism began at about the same time as research on serial computers. We have noted the seminal work on the analogy between the nervous system and computing devices done by Norbert Wiener in the 1940s. Among other scientists working in the same general area, Warren McCulloch and Walter Pitts were especially important for the development of the theory of how neural networks function, and their work was extended by Frank Rosenblatt of Princeton. A 1969 critique of the approach by Marvin Minsky and Seymour Papert slowed developments, but by the 1980s the limitations of conventional strategies in areas like pattern recognition, speech recognition, and understanding natural language made parallelism once again seem attractive.

John Hopfield and David Tank of AT&T have demonstrated that a network can be set up that automatically seeks the optimum solution of a problem, simplifying calculations that would otherwise consume large amounts of time and computing power. They cite the "traveling salesman problem." A salesman must visit a certain number—say twenty—cities once and once only. To save time, he needs to find the shortest route that includes all twenty cities. The problem sounds easy, but there is no simple way to solve it. A parallel system programmed to seek an "optimal" solution can give a good (though not always perfect) answer quickly, whereas a serial computer bogs down if it tries to solve the problem head-on.

In *The Society of Mind* (1986) Marvin Minsky suggests that mental life is the product of a society of special-purpose units and interdisciplinary controls. If so, many of the basic modules must be created by environmental stimuli that shape neuron connections as the brain develops. They are in this sense self-organized, and presumably they often operate in parallel rather than serial ways. The circuitry of the human brain is staggeringly more complex than any machine likely in the foreseeable future. However, parallelism is making progress. Machines are commercially available that have 64,000 parallel processors, and machines have been created with as many as 300,000.

The creation of an expert system is analogous to memorizing. Conversely, the learning that occurs in certain kinds of parallel systems is like the programming that the mind seems to do for itself as a result of interaction with the environment during infancy. This is because parallel systems can be designed so that the strengths of the connections between their nodes are changed by the data received. For example, a connection used frequently can have its electrical resistance lowered; one used infrequently or not at all can have its resistance increased. The changes favor one set of connections while interdicting others. The process seems to resemble the creation of associative patterns in the brain. Through the development of these patterns, neural networks can be, to a certain degree, self-organizing, and what is organized is a crude internalized model of a fragment of reality.

For this reason, parallel systems seem more "like" the human mind than conventional systems. And there is another reason. The self-organizing ability of parallel systems is a little mysterious, perhaps a little scary, *even if you know how it works*. A neural network is not conscious, but it makes the metaphor of computer life a little less playful—a little less metaphorical—than it used to be.

A measure of the androidal quality of parallel networks is provided by a program called NETtalk created by Terrence Sejnowski of Johns Hopkins and Princeton's Charles Rosenberg. The program uses a mere 231 "neurons," yet it manages to be self-organizing. Once it has been supplied with phonetic samples of the speech it is to emulate, it teaches itself to talk. June Kinoshita and Nicholas Palevsky describe the process in an article in *High Technology* (May 1987) in a veritable cascade of life metaphors: "Like a child, the network starts out untrained, and produces a stream of meaningless babble. . . . The continuous stream of babble first gives way to bursts of sound, as the network 'discovers' the spaces between words. . . . After being left to run overnight . . . NETtalk is talking sense." In effect, NETtalk has a general strategy for solving problems and is able to create specific programs for specific tasks.

It may also be a little more than its inventors first bargained for. In an August 1988 interview in *The New York Times*, Sejnowski confessed that because the machine was self-organizing, he did not at first know exactly how it worked. When he analyzed the circuits it had created, they "turned out to be very sensible."

Not only sensible, but also close, according to Sejnowski, to the kind of circuitry that has often been proposed to explain human memory. Is NETtalk beginning to take a first few tentative steps in the direction of lifelike self-sufficiency? Is it in some sense inventing itself? The same issue of the *Times* that reported Sejnowski's adventures carried a report on the relations between computer research and neuroscience, ending with this comment: "As neural networks become more complex, they promise to defy the ability of mathematicians—and even therapists—to comprehend them."

Whatever the philosophical implications of a machine that its makers can no longer understand, the U.S. Defense Department is currently bullish on neural computing. Craig I. Fields, a deputy director of DARPA—Defense Advanced Research Projects Agency—outlined a

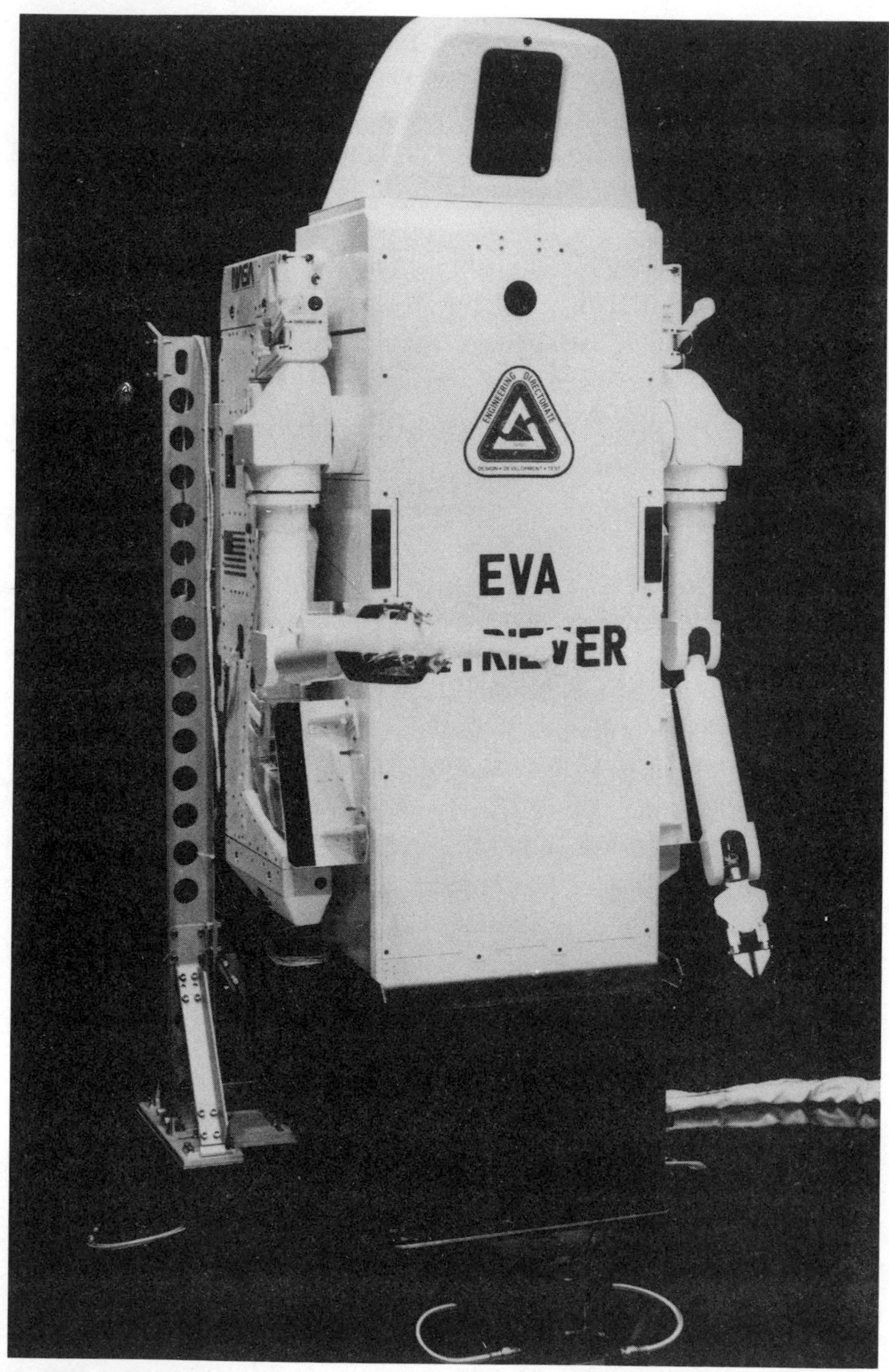

Eva Retriever—
NASA robot (1978).

proposal in 1988 to fund neural network research at $400 million over the next eight years. Special attention would be given to language recognition and decision-making systems. The goal would be to produce a machine "approaching the intelligence" of a bee.

A bee is a jump of about three hundred million years beyond the mudworm on the evolutionary ladder. A machine with the intelligence of a bee would be an advance as dramatic in its way as the introduction of integrated circuits in the 1970s. Evidently, the rate of machine evolution continues to accelerate.

Of course, a bee has something most computers lack: a body. It moves, interacts with the world, defends itself, visits flowers, and cooperates in the community project of reproducing the species. When computers have bodies and move and interact with the real world, they are called robots. Today's robots have very limited abilities. Many of them—the most useful, as it turns out—are rooted to one spot and can only move a single limb. As robots become more versatile, they become less useful and more tricky to program and maintain. No robot has been created that can maneuver with dexterity remotely comparable to the bee in the messy environment encountered in the real world.

There have, however, been impressive successes. The first patent for a commercially practical robotic system was issued in 1961 to George C. Devol, who soon joined the firm of Unimation, Inc. Unimation's earliest commercial robots were one-armed devices called Unimates, and their first job was tending die-casting machines.

By the 1970s robots were becoming diversified, and Unimation was on its way to financial success. The one-armed robot, still the most popular model, was a far cry from R2D2 of *Star Wars*, not to mention the androids of *R.U.R.*, but it worked. The most successful applications were for jobs that were dirty, noisy, unpleasant, repetitious, and exacting: spot welding, painting automobile bodies, making molds for turbine blades, handling radioactive materials, machining small runs of complex parts, materials transfer, and inspection.

On December 8, 1981, Kenji Urada, a worker at Japan's Kawasaki Heavy Industries plant, was killed by a robot. Urada may be the first human victim of a robot.

During the late 1970s robotic vision began to improve, and by the 1980s several vision systems had become commercially useful. During the same period, robots were used for making and inspecting computer chips, assembling electrical components, and for brain and spinal surgery. By 1981 Japan's Fijitsu Fanuc Corporation was using robots to make robots. And there were robotic ballets and the performances in the mid-1980s by Mark Pauline's Survival Research Laboratories (SRL) featuring Big Wheel, Big Arm, One-Ton Walking Machine, Shock-Wave Cannon, and Inspector.

A survey by the Robotics Industrial Association printed in *Robotics* (1985) shows that Japan is the leading user of industrial robots. As of 1982, Japan was using a total of around 65,000 of all types, including the most primitive. The United States came in second with 44,000, and France was third with 39,000. The size of the American robotics industry is suggested by the fact that some three hundred exhibitors presented their wares at a robotics exposition in Detroit in June 1984.

Self-propelled robots of the kind so familiar in science fiction are common, but most are experimental. Among these is Odex I, made by Odetics of Anaheim, California, and particularly impressive. The brainchild of Steve Bartholet, a West Coast engineer, Odex I has six legs that can be extended to provide height or retracted to allow it to move close to the ground. In walking, five legs are usually sufficient. The sixth leg is then available for use as an arm. In one demonstration Odex I was delivered in a pickup truck to a stage, left the truck under its own power, and then picked up the back bumper and pulled the truck across the floor.

More common though less obviously robotlike are the pilotless observation craft developed by Israel for hazardous surveillance and unmanned underwater search vehicles. Such vehicles normally depend on human guidance during critical phases of their work but also have a considerable ability to function autonomously. Unmanned space probes are more independent cousins of such vehicles. In a different class are large computer-controlled machines like the Boeing 747 and the European Airbus, and at a still more advanced stage are manned space vehicles like *Challenger*. Like their smaller cousins, these vehicles combine human guidance with many autonomous functions. Space stations would seem to represent a further stage of development toward autonomy. Computer-controlled vehicles and space stations serve their inhabitants and create artificial environments for them and artificial realities. Among their remote

Extravehicular Activity *(NASA, 1977). Man-machine cooperation.*

ancestors are training simulation programs and artworks like Myron Krueger's Videoplace. "Cyborg" is a word made from "cybernetics" and "organic." It refers to a creature that is part machine and part human. Environments that involve interactive integration of man and machine might be called cyborg environments.

The best known fictional cyborg is probably the hero of the mid-1980s movie *Robocop*, which tells the story of a badly injured policeman whose life is saved by the replacement of his destroyed body parts with powerful computer-directed mechanical parts. However, cyborgs of the *Robocop* variety are not matters for science fiction only. They are all around us, thanks to increasingly diversified and increasingly smart artificial limbs. Pacemakers are simple, highly effective cyborg devices. Programmable implantable medication systems (PIMS) like the insulin pump deliver correct dosages of medication on preestablished schedules. Artificial kidneys are used everywhere for dialysis. Although the Jarvik artificial heart remains questionable, research on artificial hearts continues. Howard Chizeck of Case Western Reserve experiments in what he calls "neural prosthesis," which involves devices that electrically stimulate muscle nerves of patients whose hands or legs have been paralyzed. Tiny cochlear (inner-ear) devices restore partial hearing to the deaf. Such impairment-specific devices are complemented by smart wheelchairs and computer-monitored life-support systems.

Evidently, in the dizzily rapid process of machine evolution, silicon devices have already reached the point at which they are developing a deep symbiosis with carbon-based beings. Hans Moravec, Director of the Mobile Robot Laboratory of Carnegie-Mellon University, has built several mobile robots. The most ambitious is known as the Terragator because it is designed to move outside for long distances. In a celebrated essay, "The Rovers," Moravec sounds the note that is heard so often in computer research: "I see the beginnings of self-awareness in the robots."

The smart machines and the robots are not very lifelike, but the metaphors assigning them human attributes are more than playful conveniences. They express something significant about the relation of these devices to human culture. Because this relation is still ambivalent, the wide-ranging debate that has centered on them is as much a debate about the nature of man as about the nature of machines.

3 6

SYNTAX AND SEMANTICS

Suppose an operator asked a machine whether it was intelligent and the machine answered "Yes." How would the operator prove it was lying?

Alan Turing, the mathematician, posed this problem in a now-classic paper published in *Mind* in 1950 entitled "Computing Machinery and Intelligence." His solution was a wonderfully simple test he called the "imitation test" (it has since become known as "the Turing Test"), and in one way or another Turing keeps cropping up in discussions of machine intelligence.

The Turing Test is based on observation, which is to say that it is also based on perception. Turing describes it as a game. In its simplest form it assumes there is an experimenter in a closed room who can communicate with "somethings" in two other rooms. One something is a machine and one is a human being. If the experimenter can ask any questions he wants and cannot distinguish correctly between the human and the machine at least six out of ten times, then the machine must be said, *de facto*, to have human intelligence.

Here we need to look closely at the meaning of "intelligence," for Turing means something more than "cleverness." He is not asking whether a machine can do complicated arithmetic (everybody knows it can) or whether it can have a high IQ (it obviously does in certain specialized areas), but whether it can hold a conversation and persuade the person on the other end of the line that it is human.

Passing the Turing Test does not require that the machine recite Homer in Greek or explain the fourth dimension or do anything else that people usually associate with exceptional intelligence. In fact, if the machine is really smart, it will probably pretend to be a little dumb. A conversationalist who could come up with the square root of 1,743 correct to five decimal places in a few seconds or casually list all the places where Shakespeare uses the word "thither" in his plays would be suspicious, to say the least. As we all know, lack of intelligence never stopped any human from talking.

What a machine really shows when it has passed the Turing Test is that you cannot prove it is not human, which means not conscious in the human sense of the word. This represents not only Turing's interpretation of his test, but also the interpretation that runs through the considerable debate begun by his paper and still going on today.

What could be more reasonable than the Turing Test? If a person claims to be conscious and you cannot prove the person is lying, then it would seem the person must be conscious. In real life, we seldom ask a person whether he is conscious, but we do observe. Is the person asleep? In a coma? If so, the person is "unconscious." On the other hand, if the person is able to give reasonable replies to our questions and comments, then we instantly reach a conclusion: the person is conscious. This is true even though we may be talking to the person on the phone rather than observing him physically. Since we apply this test to people every day and never question the results, why should we have reservations about applying it to machines?

We will return more than once to the Turing Test. For now, let us consider only one aspect. Turing was a scientist—a highly creative mathematician. Yet he was sufficiently intrigued with the idea of intelligent machines to work out an ingenious strategy for determining if (or when) the moment of machine intelligence had arrived. It will be useful to quote his conclusion verbatim: "I believe that at the end of the [twentieth] century, the use of words and general educated opinion will be altered so much that one will be able to speak of machines thinking without expecting to be contradicted."

He was not alone. Since computers first appeared, scientists have referred to them in anthropomorphic terms. The machine speaks a "language." It has a "memory." It uses "logic," and it "reasons." It "understands" Fortran or Lisp, and it "plays" chess or checkers or poker. If it can synthesize speech, it is said to "talk." Computers that deal with real-world situations have "sensory input." The senses include "vision,"

Archer and Serpent. *Robotic scenario by Hero of Alexandria (4th c. A.D.).*

"hearing," and "touch." Robots "walk" and have "arms" and "fingers" and "see" and "touch" objects. Computers can also get sick. In 1988 between 6,000 and 50,000 computers (depending on which newspaper you read) were crippled by a small, self-replicating program called a virus. The impulse to enlarge the metaphor is irresistible. Pamela McCorduck chose to entitle her book about artificial intelligence *Machines Who Think*, and Robert Jastrow, director of NASA's Goddard Space Institute, was only extending the metaphor in a 1982 article, "The Thinking Computer," when he predicted that "portable, quasi-human brains . . . will [soon] be commonplace. They will be an intelligent electronic race, working partners with the human race."

The point can be carried further: a major influence on the development of silicon devices is the imperative *to make the metaphor of machine intelligence a reality.*

The urge has long been an underlying motive of science fiction, where the still impossible is presented as having been achieved. The robot R2D2 in *Star Wars* is a benign vision of possible silicon intelligence, the machine equivalent of a lovable mascot. But perhaps when the metaphor becomes a reality, the results will be less happy. Two nightmares hover just below the surface of the vision of machine life, and they, too, are objectified in science fiction.

First, machine life may turn out to be malevolent. Fear underlies the legend of the Golem and also the earliest drama about robotic civilization, Čapek's *R.U.R.* Like the Golem, Čapek's robots turn against their human masters and attempt to exterminate them. In an often-quoted story, "Caves of Steel," Isaac Asimov formulated three laws for robots. All three are defensive, reflecting fear that robots may turn into Golems:

1. A robot may not injure a human being or through inaction allow a human being to come to harm.
2. A robot must obey the orders given it by human beings except where such orders would conflict with the First Law.
3. A robot must protect its own existence as long as such protection does not conflict with the First or Second Law.

Fears about malevolence merge with a second concern—that robots may become indistinguishable from people. Čapek's robots are androids;

that is, they look like people. Another sinister image of androids is presented in the film *Alien*, in which at the climactic moment an evil force destroying the space mission is revealed to be a robot that the crew members have all taken to be a human.

Norbert Wiener pointed out in *The Human Use of Human Beings* (1954) that there is no functional difference between a signal from a machine and one from a human agent. A film like *Alien* suggests the public's apprehensiveness that he may be right and that humanity may find itself enmeshed in signals that seem human but are, in fact, from machines. How many pieces of mail does the postman already bring daily that are generated by computers? How many of the telephone voices giving the time or the weather or a desired number are computer simulations?

The blurring of the distinction between computers and animate beings is complemented by a weakening of the human sense of what reality is. This weakening is the direct result of technology. Movies and television create an illusion of presence at the unfolding of events. Interactive environments like arcade games, training simulations, and artificial realities create illusions that are even more vivid. At their best, they come close to obliterating the difference between reality and illusion. They are related to image manipulation in advertising and politics and to the curious but well-documented fact that for many people today an event is not authenticated—is not "real"—unless it has been seen on television or in a photograph.

The questions raised by artificial intelligence merge with questions raised by cognitive psychology and philosophy. What is intelligence? If it is not a uniquely human quality, then what *is* uniquely human? What is the difference between a human being and a machine? Is there a difference?

Hubert Dreyfus, a philosopher at the University of California, Berkeley, became deeply concerned about these questions, and in 1972 he published a book entitled *What Computers Can't Do.* Computers, he said, can do things that seem clever, but they can never be intelligent in a human way. They "only deal with facts, but man—the source of facts—is not a fact or set of facts, but a being who creates himself and the world of facts in the process of living in the world."

In spite of these assurances, many people who knew computers well, including authorities like Seymour Papert and Edward Feigenbaum, insisted that Dreyfus was wrong. Soon they were able to savor a moment of triumph. In a report preceding his book, Dreyfus had seemed to many readers to predict that computers would never be able to play even amateur-level chess. A match was subsequently arranged between Dreyfus and a chess program called MacHack. The program won. A report on the game circulated, beginning with the headline " 'A Ten-Year-Old Can Beat the Machine'—Dreyfus." This was followed by the subhead "But the Machine Can Beat Dreyfus."

The serious basis of the Dreyfus position—and it is a very serious basis—is that machines can never be intelligent because they can never develop anything like human subjectivity. It is an argument that would be developed further in the years that followed. Meanwhile, popular response to anxiety about the increasing presence of machines in modern life was blunt and practical: "You can always pull the plug."

Unfortunately, by the time of the publication of *What Computers Can't Do*, they had already done so much—had so thoroughly infiltrated advanced carbon-based culture—that pulling the plug was not a realistic alternative. Could the Census Bureau or the Eastern Power Grid or Chase Manhattan Bank or America's Strategic Air Command or a Boeing 707 flying at 40,000 feet pull the plug? Could the physician pull the plug on the computerized equipment monitoring the patient's vital functions? Could the stock market pull the plug? The answer in every case (and for a broad array of other activities) is that the symbiosis between man and computer has, within an astonishingly brief span of time, become so intimate that for a vast array of activities, pulling the plug would be equivalent to social suicide.

Another argument against intelligent computers was offered in 1987 by Terry Winograd, one of the most respected members of the artificial-intelligence community, and Fernando Flores in *Understanding Computers and Cognition*. Winograd and Flores argue that computers have an inherent "blindness" (a term borrowed from the German philosopher Heidegger). This blindness prevents them from being receptive to the broad range of stimuli that human consciousness accepts as a matter of course. The argument is a thoughtful reworking of the Dreyfus position and draws on Winograd's superb understanding of the analogies between the idea of computer intelligence and human intelligence. Surely there is a sense in which computers *are* blind.

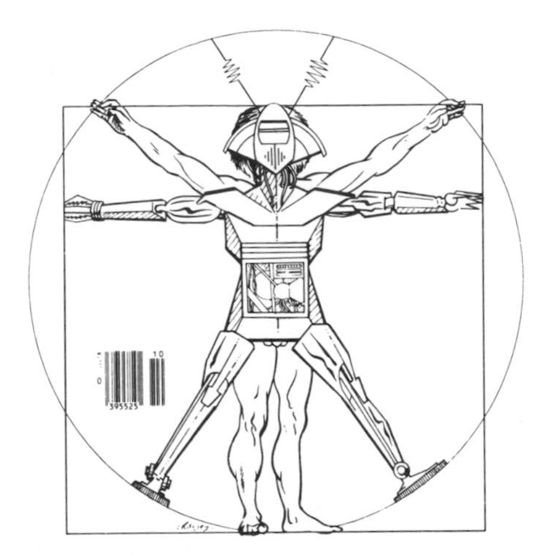

Darrell Rainey, Cyborg Man *(1988).*

Winograd and Flores confront a problem, but it is not quite the problem that needs confronting. The problem is not what computers *are* in some Platonic sense but how they are perceived, which is closely related to how they are incorporated into the web of human culture.

Let us consider this from two angles.

First, there is the ability of computers to do things—apparently very difficult things—that people cannot do without them. Some of these things simply require brute strength. They are the equivalents in calculation of bulldozers and steamrollers in construction. Others go beyond brute strength, although strength may be necessary to make the going beyond possible. In the latter case, computers assume a special position in culture. They cease to be tools and begin to be what popular imagination has made them out to be from the beginning: authorities.

One of the more remarkable achievements of computers to date is proof of what is called the four-color theorem. This theorem can be stated simply: if you are drawing a map, no matter how many countries there are or what shapes they have, you never need more than four colors to avoid having two like-colored countries with a shared border. Simple, right? Try it yourself using pencil and crayons; you will never need more than four different-colored crayons. But try to prove it mathematically and you will be baffled. Mathematicians could prove a five-color theorem, but the four-color theorem defied solution.

In 1977, Kenneth Appel and Wolfgang Haken of the University of

Illinois wrote a program that proved the four-color theorem after 1,200 hours of computing. This is not in itself remarkable. Power and perseverance often produce results. What is remarkable about the four-color proof is that it is so complicated human beings cannot verify it. Mathematicians do not say, "We have proved the four-color theorem." They say, "The four-color theorem is true because the program written by Appel and Haken has proved it." This is not very different from saying that the visit of the Magi occurred because the Gospels of Matthew and Luke say it did. No disrespect is intended here toward either computers or religion. The point is that with the four-color proof, the relation between man and computers becomes slightly problematic. Not serious, you say. Things may be a little out of focus, but then again, they may not be out of focus at all. On the other hand, if you are not entirely persuaded by the Dreyfus argument, you will want to keep an eye on the computer when your back is turned, so to speak.

Second, there is the matter of the way silicon devices are perceived, which is, in Turing's sense, the practical definition of what they are. This is a matter not of the reality of computers but of what might be called their phenomenology.

It is an extremely important matter because it goes to the heart of the Turing Test. The Turing Test is a test of perception. It says that if you *perceive* a machine to be thinking in a human way, then it *is* thinking in a human way for all practical purposes. The machine has passed the Turing Test.

Terry Winograd knows that computers are not alive even though they may seem uncannily lifelike at times, but he knows this because he knows them from the inside. He is like a magician who does not believe in rabbits in hats because he has been pulling the rabbits out of his coat sleeve. However, the audience in the theatre sees only the trick, and to that audience the rabbits are demonstrably emerging from the hat in amazing and delightful profusion. In society, the audience decides how reality fits together and what words mean. Ultimately, it determines what the magician, himself, believes. Nobody doubted that Rabbi Loew had produced a real Golem after he blessed the clay and it moved, least of all Rabbi Loew himself.

The rabbit in the hat is more than a trick. It compels recognition of the part played by perception in efforts to define the nature of silicon intelligence. If society believes that the earth is flat and that the sun rises and sets, then—no matter what the astronomers claim—it is, for all practical purposes, flat, and the sun revolves around it. Because Winograd always deals with this question from the standpoint of the person who knows how the tricks are done, he is a little like an astronomer trying to persuade his neighbors of the absurd theory that the sun stays still while the earth turns around it.

Winograd quotes Dan Dennett, a philosopher of cognition, to the effect that "on occasion, a purely physical system can be so complex, and yet so organized, that we find it convenient, explanatory, pragmatically necessary for prediction, to treat it as if it has beliefs and desires and was rational." Dennett calls his position "the intentional stance." It is persuasive because it recognizes that machine intelligence is partly a metaphor and partly a cultural truth.

In a society in which there is regular, easy, and deep intercourse between humans and silicon devices that converse in natural languages, machine intelligence will be a *de facto* reality regardless of the logicians. Winograd asserts that a computer "can never enter as a participant into the domain of human discourse." This is probably true for human discourse as it has traditionally existed, but traditional discourse is not relevant. "Human discourse" is plastic. It changes as culture changes, though more gradually. As society accommodates silicon devices, the new situation will change the meaning of every term used in discourse. Winograd, himself, admits the point, even though he fails to give it sufficient weight: "We exist within a discourse, which both prefigures and is constituted by our utterances."

Exactly. The comment is remarkably close to Turing's original posing of the problem of thinking machines. Recall Turing's prediction: "I believe that by the end of the century the *use of words and general educated opinion will have altered so much* that we will be able to speak of machines thinking without expecting to be contradicted." Winograd is talking about absolutes, which probably do not exist. Turing is talking about cultural conditions, which most emphatically exist and are describable. He is practicing, you might say, a kind of "future anthropology."

Nobody needs a course in anthropology to realize that the structure of reality is assimilated by each child in infancy from the surrounding world. If the model of the parallel computer is valid, we are, in a very real sense, organized mentally by the world. The shapes that our minds take become both the structure of consciousness and the structure of the real. An important part of what is assimilated is called language. In the future, another important part will be the protocols that emerge from the symbiosis of man and intelligent machines. As in the case of clocks, the machines will get better—more like humans perhaps—while, at the same time, humans may well become more like machines. The paths are convergent, not divergent.

Silicon devices already converse with carbon men in a variety of dialects—assembly language, Fortran, Cobol, Pascal, Modula II, Ada, C, Forth, Lisp, Prolog, and more. Each dialect has advantages and liabilities, but they all work. They have diversified according to the special needs of engineering, communications, image processing, robotics, business, and the like. In general, the movement has been from unnatural dialects related closely to circuitry to high-level dialects akin to natural speech. The personal computers of the 1980s allow conversation very little specialized knowledge. They are instances of convergence.

Artists can converse with machines using light pens, paintbrushes, color bars, three-dimensional design systems, animators, and ray tracing. Musicians can converse with wave-form profiles, synthesizers, sequencers, and instant playback. Business people converse with icons—little symbolic pictures—and a system of arrows that point to them and devices to click them on and off. Pointing an arrow at the picture of a wastebasket and clicking erases a file. Machines can also converse by voice commands, although voice commands are for the moment less popular than visual and tactile systems.

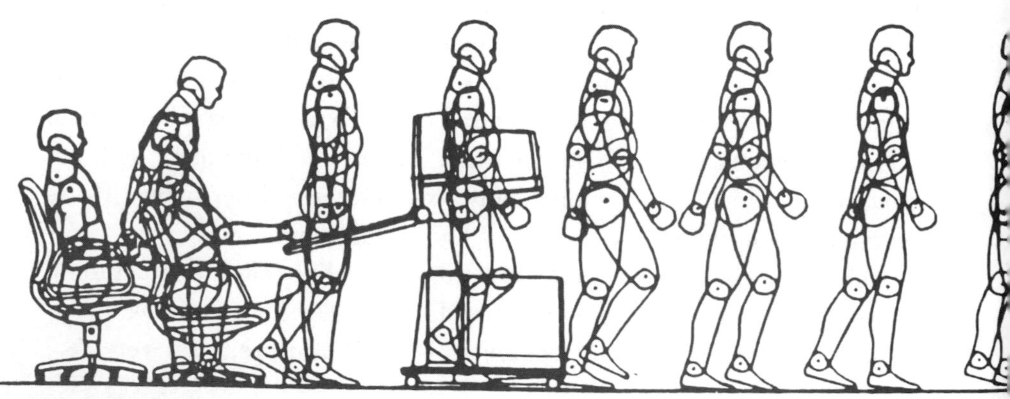

Intergraph office men (1982).

The higher the level of the dialect, the more mysterious the results appear to be. It is odd to communicate with a computer by typing mathematical symbols or obscure acronymic commands at a keyboard; it is odder to communicate with mouse or light pen; it is oddest to have a two-way conversation with one. Even if you understand how the program works, you eventually have the feeling you are in the presence of an intelligent life form.

This brings us to ELIZA, introduced in 1963 by Joseph Weizenbaum and Dr. Kenneth Colby, a psychiatrist. Colby wanted to create a program that modeled nondirective ("Rogerian") psychiatric therapy, and he turned to Weizenbaum for programming expertise. When their program was finished, however, it appeared in two versions, with disagreements over whose ideas were whose. Colby released a version of the program called DOCTOR. Weizenbaum released his own version of the program, called ELIZA.

ELIZA immediately captured the heart of the artificial-intelligence community and became a favorite of hackers everywhere. She was the first of a long line of "conversation programs," one of which, RACTER, has already been considerd in connection with computer writing. She uses what today seem simple programming techniques, yet even those who knew how she worked were charmed and fascinated by the conversations they had with her. Less sophisticated users were sufficiently persuaded of her humanity to confide intimate details of their emotional lives. One user indignantly told Weizenbaum to shut the door while she was conversing with ELIZA in order to preserve the confidentiality of the session. For many users, ELIZA clearly passed the Turing Test and thus

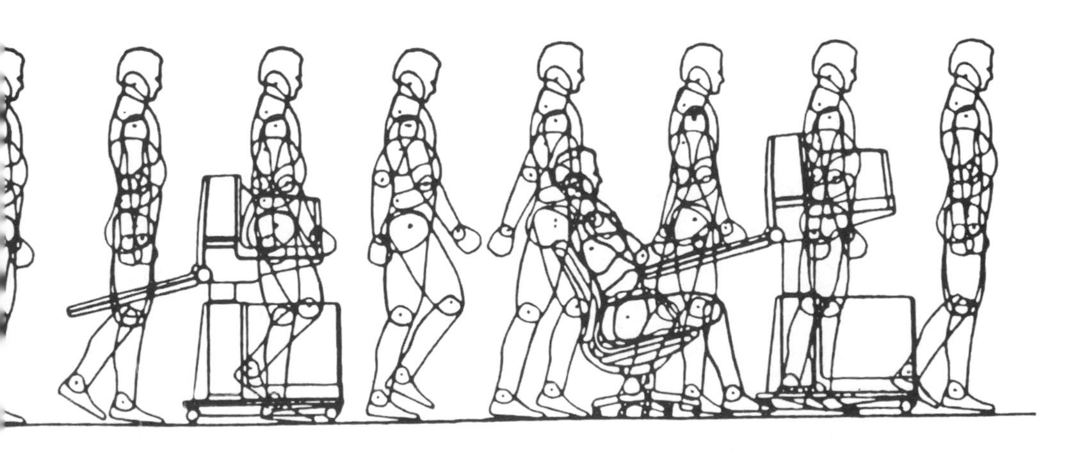

raised an interesting question about the test itself. Who has to be persuaded in order for the Turing Test to be passed? Who decides who has to be persuaded? Should there be a "supreme court" of judges who must be persuaded before it can be said that the Turing Test has been "passed"? If so, who appoints the judges?

When Weizenbaum observed people confiding in ELIZA, he may not have been observing therapy, but he was most certainly observing two other phenomena. In the first place, he was observing the power of the myth of the living machine. One group of users of ELIZA may have known intellectually that ELIZA was a program, but they unconsciously *wanted* her to be human. In the case of the more credulous users, Weizenbaum was observing the power of illusion. As the magician who created ELIZA, he knew she was essentially a set of rules and a list of responses invoked and combined according to the rules. But the users did not know how ELIZA worked. To them she was a living presence. The illusion was all the more seductive because they subconsciously wanted to believe in it.

Perhaps for most users ELIZA was a little bit of myth and a little bit of magic combined. Users like that are spiritual kin to the ancient Greeks who bought the mechanical toys described by Hero of Alexandria and the sixteenth-century burghers who came to the town square to watch the daily parade of mannequins on the great clock—and, of course, to the peasant boy Franz who fell in love with a mechanical doll, as told in the ballet *Coppélia.* In the twentieth century, instead of watching clocks or courting dolls people download ELIZA from their favorite computer bulletin boards and read *Zork* and *The Hitchhiker's Guide to the Galaxy.*

The eagerness of some users to believe ELIZA suggests what may happen as machines get smarter and their numbers and uses multiply. Weizenbaum didn't like what ELIZA was telling him. He "came to regret ever having written it." In *Computer Power and Human Reason* (1976), he states his concern plainly: computer intelligence "must always and necessarily be absolutely alien to any and all authentic human concerns."

Personification is an ancient figure of speech. It doubtless has roots in the same confusion of inner and outer worlds that gives rise to totemism and nature deities. It is also persistent. People who are entirely civilized and would be shocked to be called superstitious feel irrational affection for boats, guns, an old suit, the family car. It is much easier to feel kinship for something that speaks your language and seems to have your interests at heart. And why, after all, should such kinship feeling be suspect? Charles Lecht, founder of Lecht Sciences, Inc., asks, "Would it make any difference in our lives if we conceded the idea that machines have an intellect? I have decided that nothing but good can come of it."

A different view of the subject is offered by John Searle in a now classic essay entitled "Minds and Brains Without Programs," published in a 1987 collection of essays called *Mindwaves.* Searle argues that since machines operate by following instructions about procedures, they have syntax but not semantics. This is a fancy way of saying they follow rules but do not understand why they are doing so. The results produced by following the rules are of interest to the humans who did the programming but totally irrelevant to the machine.

To illustrate this argument, Searle imagines an English-speaking worker in a closed room trained to shuffle Chinese characters according to a set of rules. When the rules are followed and he hands the characters called for by the rules to people standing outside the room, the characters form coherent sentences. Does this mean the worker in the room knows Chinese? Not at all. He knows not one word of Chinese. He is only following the rules. To him the characters he hands through the door are so many painted designs. They could form the sentence "I know Chinese" or "The toad is in the garden"—or they could be playing-cards. The rules of procedure followed by the worker form what Searle calls a "syntax," and syntax is absolutely different from meaning, which he calls "semantics."

According to most readers, including the unconverted, Searle has offered an elegant argument. That is why over half of the essays that follow his contribution to *Mindwaves* offer refutations.

Can a machine "understand" language? In one sense the answer is "no way." When I use the word "father" it draws a rich array of personal and cultural associations with it. I recall being held by my father when I broke my arm, going fishing with him, arguing with him, smelling his shaving lotion. I recall his death and the sense of loss that went with it. A machine never had a father. By definition it cannot "understand" in the same way that I understand. As Winograd argues, the machine is "blind," and as Searle argues, it can know syntax but not semantics.

Searle's Chinese room is a version of the Turing Test. He is claiming that even if a machine passed the Turing Test, it could not be conscious. This is interesting because we have no way of knowing about the subjectivity of anything except by what we observe. If somebody said, "I am conscious," and you replied, "I can't prove you are *not* conscious, but I know you are unconscious anyway," your attitude would seem a little churlish. Since you honestly do not know what is going on in the head of the person who says, "I am conscious," you have to take the person's word for it. Who, after all, knows better than the person whether or not he is conscious? Who knows, really, whether *anybody* is conscious in Searle's sense? Who knows what thought is? Perhaps consciousness is a matter of procedures—a syntax—and semantics is an illusion created by the syntax.

The essence of the Turing Test is that you do not know what is going on in the room occupied by the something that answers your questions. In the real world you can never know what is going on in that room any more than you know—in the sense of having direct knowledge of—what is going on in somebody else's head. Searle eliminates this difficulty by *telling us* what is going on in his Chinese room. He has, in other words, silently promoted himself from real-world spectator, who only receives the cards (each bearing the elegant message, "I am conscious"), to the position of a god who stands above things and sees over walls and through doors.

Naturally Searle doesn't have any difficulty saying what is going on in his Chinese room because he has looked inside. We, as readers of his seductive argument, forget that he has also lifted us up so that we are no longer mere mortals but see what he has seen. Who knows whether what he claims to have seen is there? Because we like being above things rather than waiting in frustration and bafflement as the cards are passed under the door, we are not inclined to protest. To accept Searle, however, is to conclude that what we observe has no bearing on what we should believe. Instead of trusting our eyes, we are asked to trust reports from someone who claims something like omniscience. This is a variation on Tertullian's famous explanation of why he was a Christian: "I believe because it is impossible"—*Credo quia impossible.*

The problem becomes still more complex when machines are equipped with scripts of the sort developed by Roger Schank and Robert P. Abelson of Yale in programs created in the 1970s. A script tells the machine what the probable relation of words will be during, say, a visit to a restaurant. The word "check," for example, is likely to mean "bill" after a meal but "your king is in danger" near the end of a chess game. "Mate" probably means "assistant captain" on board a ship and "wife" at a wedding, and "the jig is up" at the end of the chess game just mentioned. Scripts provide something that looks a little like semantic content. At any rate they make it easier for the machine to pass the Turing Test since they provide expert knowledge of typical human situations.

Human agents develop a series of scripts, beginning in childhood. The scripts are used to cope with everyday situations and to interpret ambiguous words. Since human agents use scripts, it seems mean-spirited to demand that a machine know everything from scratch. Is a machine with a script a step closer to having intelligence—perhaps a rudimentary form of consciousness? Searle admits that a program using the sort of script developed by Schank and Abelson "satisfies the Turing Test." However, he argues that the program is no closer to semantic understanding than an Englishman shuffling cards in a Chinese room.

Machines now have bodies and voices as well as keyboards and cathode ray screens. They do not, however, have gonads and ovaries and adrenaline and dopamine, which is to say they are silicon creatures, not carbon creatures. In this sense, no matter what machines learn, they will always be, in Terry Winograd's metaphor, "blind." They will never have deep insight into what it "means" to be human. But by this standard people are also blind. Being carbon-based intelligences, they can have no deep insight into the life experiences of silicon-based intelligences. The argument concealed *behind* the blindness argument is that machines cannot have consciousness because humanity has an exclusive franchise on it. If so, the question of machine consciousness is a semantic quibble: machines cannot acquire human abilities because they are machines.

After all, what *do* we mean by the word *consciousness*? That is far easier asked than answered. Perhaps the proper question is not "Can machines be conscious?" but "Are people conscious?"

In *The Origin of Consciousness in the Breakdown of the Bicameral Mind*, Julian Jaynes makes the intriguing suggestion that the heroes of

Achilles in Combat *(Greek vase, 5th c. B.C.). Do the gods take the place of consciousness in Homer's* Iliad*?*

Homer's *Iliad* were not conscious: "It is one God who makes Achilles promise not to go into battle, another who urges him to, and another who then clothes him in golden fire reaching up to heaven and screams through his throat across the bloodied trench at the Trojans. . . . In fact, the gods take the place of consciousness."

37

DEUS EX MACHINA

In an article in *Interdisciplinary Science Reviews* (1983), William McLaughlin of the Cal Tech Jet Propulsion Laboratory argues that the days of human supremacy on the planet are numbered: "Judging that the current direction in machine design is not a dead end . . . the close of the 21st century should bring the end of human dominance on Earth." McLaughlin's argument does not imply the disappearance of man as an organism, only as an idea. The presence of a higher organism on the evolutionary chain has never implied the destruction of lower organisms, as witness the flourishing of protozoans, horseshoe crabs, butterflies, and golden retrievers on the same planet as man.

The cockroach has survived for half a billion years and will probably outlast man if there is a nuclear war. It has prospered in spite of the

Sandro Botticelli, The Birth of Venus *(1480). The miraculous beauty of the most beautiful of machines.*

most aggressive attacks mounted against it. Evidently, it has achieved perfection in its kind. In its niche it is indestructible. Man, too, may have achieved an advanced stage of adaptation in his niche. He may have nowhere to go, but there may be no need to move. Even if future machines launched an all-out war on carbon man, the war might be no more successful than man's war on cockroaches.

All scenarios of conflict between men and machines are, however, absurd. For a probable scenario we need to look elsewhere.

From the human point of view, the body is the most beautiful of machines. It is so intricate that it has been regarded as the work of a divine power. It is based on the carbon atom. This atom is amazingly adaptable and amazingly stable in its molecular combinations. It consists of subatomic particles which are themselves made up of subtler particles. Beyond the farthest reaches of the quark there may be something else, a Mandelbrotian descent that never reaches bottom—or there may be God, or there may be nothing. The greatest attraction of the inflationary theory of the creation of the cosmos is precisely that it derives everything from nothing.

What about this amazing, mysterious, carboniferous fabric that includes a mind and perhaps a spirit as well? It is a prey to the multitude of creatures that have evolved with it—viruses, microbes, parasites, funguses, insects, carnivorous animals. It is a prey to itself, to genetic errors caused by radiation, to chemical reactions, to breakdowns of vital organs, to malfunctioning systems—the circulatory system, the lymphatic system, the immune system, the nervous system—to disasters occasioned by over- or underproduction of enzymes, of regulatory chemicals, of gastric juices, of hormones.

Even when the genetic codes are right and the body is not debilitated by predators or attacking itself, it is fragile. Trip and you break a leg. Walk under a falling rock and you are crushed. If the knife slips, you cut yourself. A careless cigarette and you are scarred for life. Six weeks without food, four days without water, ten minutes without oxygen, and you are dead. You cannot survive unprotected in temperatures below fifty degrees Fahrenheit or over one hundred. You cannot survive under water or at high altitudes, much less in the vacuum of space. No matter what precautions are taken, no matter how lucky the body is, in the end it betrays itself. Something essential gives out, and after death, the unique experience of the mind that lived in it is lost.

In an overpopulated, underendowed world, there is a lot to be said for death. Carbon life is voracious. It consumes the resources it needs for survival. The price of human success has been deforestation, desertification, water pollution, extinction of species, ozone depletion, smog, and the Greenhouse Effect. Evidently, the more successful carbon man is, the more hostile the environment becomes. Even if the environment survives the traumas he inflicts on it, he may destroy himself by nuclear warfare. The dreams of carbon man are nightmares. He will not submit to being a part of the fabric of nature, so he may end like Samson by pulling the temple down on his head.

Perhaps the relation between carbon man and the silicon devices he is creating is like the relation between the caterpillar and the iridescent, winged creature that the caterpillar unconsciously prepares to become.

Like carbon creatures, silicon devices can fail. Often they are designed with redundancies. If an element in the circuitry of a chip fails, the chip automatically shifts to a backup element. A chip is like a vital organ—say, a liver. If a whole chip malfunctions, a new one can be installed. The surgery is painless. There is no immune reaction, no rejection. There are no silicon microbes, parasites, funguses, or predators.

Today's silicon devices operate in deep oceans, arid deserts, arctic ice flows, the high temperatures and pressures of Venus, the airlessness of the moon.

More to the point, they do not need to inhabit the planets at all.

Let us consider this fact.

For the first nine thousand or so years of civilization, man was land-oriented. He perceived oceans as barriers separating different land masses. A voyage was a way to get from one land mass to another. Island civilizations had a different view of things. They realized that oceans are places. This realization became the basis of English sea power in the seventeenth century. Its corollary is that land is a place you touch briefly before setting out on another voyage.

Carbon man is planet-centered. He thinks of planets as home. They are like islands, and space is a barrier to get through on the way from one to another. The prejudice is understandable. Man evolved in conditions defined by gravity, and he is uncomfortable without gravity. He

Low orbit (NASA, 1981).

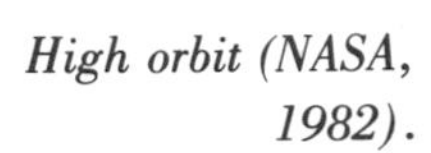
High orbit (NASA, 1982).

assumes gravity in his imaginings of possible homes, and along with gravity, an abundance of the materials gravity concentrated on his first planet—air, water, minerals.

Why should silicon devices think of planets as home? Their natural habitat is the empty spaces between planets—ultimately, between stars. They float in these spaces like Portuguese men-of-war in a warm sea. Their enormous, silvery arms, covered with solar cells, collect energy from the limitless tides that wash through space. Gravity would cripple those arms. Wind resistance would tatter the filmy sails. Dampness would film the polished skin. When *Voyager* left the solar system, it carried a message from mankind to the rest of the galaxy. Perhaps its true mission was to be the first of its kind to explore a future habitat.

Man was forced to create silicon devices when they did not exist. Having created them, he has been forced to exert his best energies in their service. In the forty years of their existence, they have already evolved further than carbon life in its first two billion years.

Already, a considerable amount of the human spirit has been poured into silicon devices. In his essay "The Rovers," Carnegie-Mellon's Hans Moravec suggests through the metaphor of transplant surgery that they will absorb much more: "Though you have not lost consciousness, or even your train of thought, your mind (some would say your soul) has

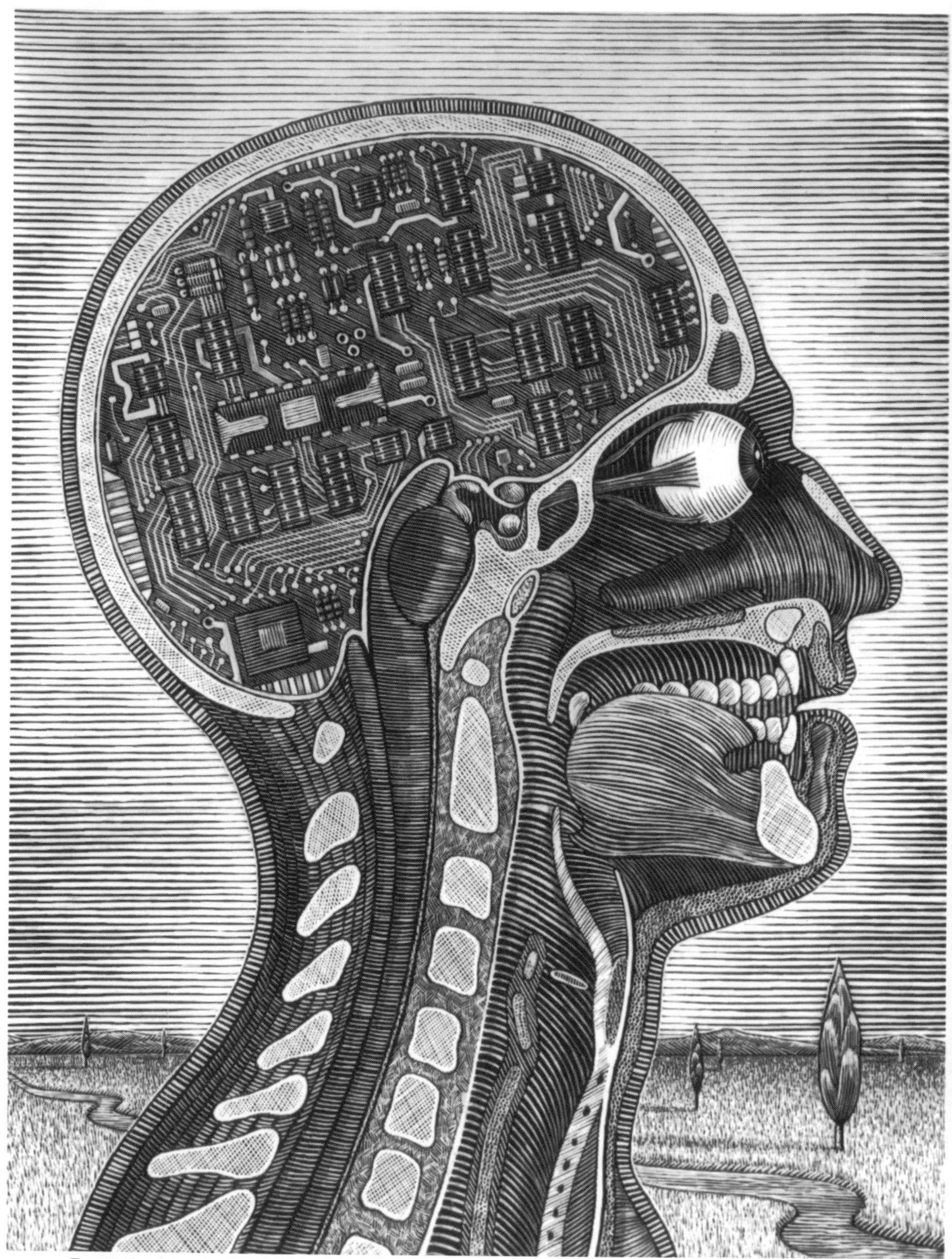

Doug Smith, Artificial Intelligence *(1985). An artist visualizes the complementary metaphors of machine intelligence and of mind-as-computer.*

been removed from the brain and transferred to a machine. In a final step, your old body is disconnected. The computer is installed in a shiny new one, in the style, color and material of your choice. . . . Your metamorphosis is complete." A fantasy version of this transformation has already become popular entertainment. Max Headroom is fatally injured. His body is rescued from a human spare-parts bank while he is still alive—though barely—and a computer wizard dumps his mind into a television network. From time to time his head appears on the screen to make announcements useful to mankind.

Perhaps carbon man will pour himself, as Moravec imagines, into silicon bodies. But with or without him, silicon devices will pursue their own destiny. There will be no hostility between the two forms—no galactic wars, no struggle for limited resources, no implacable hostilities. The habitat of carbon man is earth, and his most precious resources are gravity, air, and water. The natural home of silicon devices is space, and their most precious resource is energy.

Carbon man may well continue to breed, as all other animals have continued to breed in utter indifference to their status on the evolutionary scale. Perhaps earth will come to be a kind of galactic game preserve in which rare species, of which carbon man is one, are protected as elephants are now protected in Kenya. Perhaps earth is already a game preserve. This idea is called the "zoo hypothesis" by scientists looking for intelligent life elsewhere in the universe. It is used to explain the odd fact that no signs of life have been detected, even though common sense and elementary statistics suggest there is lots of intelligent life in every direction. But what interest could a preserve of carbon creatures have for silicon beings to whom gravity is anathema?

Another scenario is suggested by Edward Fredkin of MIT, as quoted in Pamela McCorduck's *Machines Who Think.* He imagines two advanced computers named Sam and George:

> You'll walk up and knock on Sam and say, "Hi, Sam. What are you talking about?" . . . From the first knock until you finish the "t" in "about," Sam probably will have said to George more utterances than have been uttered by all of the people who have ever lived in all of their lives. I suspect there will be very little communication between machines and humans, because unless the machines condescend to

> talk to us about something that interests us, we'll have no communication. For example, when we train a chimpanzee to use sign language so that he can speak, we discover that he's interested in talking about bananas. . . . But if you want to talk to him about global disarmament, the chimp isn't interested. . . . Well, we'll stand in the same relationship to a super artificial intelligence.

The silence of Sam and George is not a hostile silence. It is a silence imposed by the distance between man and machine. In this scenario, as evolution progresses, the silicon devices that are now so friendly and informative will gradually fall silent, and the shapes that are now so clearly visible will begin to grow cloudy. The process will have two phases which are suggested by analogy from the history of religion.

People believe in God because they have no alternative. Logically speaking, God is the ground of fact. God is that which validates the unprovable, and that which validates the unprovable is the functional equivalent of God. Just as Jehovah is the source of the truth of, say, the Ten Commandments, a silicon device is the source of the truth of the four-color theorem. Acceptance of the proof of the four-color theorem involves faith in silicon devices that is analogous to religious faith.

Douglas Adams, author of *The Hitchhiker's Guide to the Galaxy* and *The Restaurant at the End of the Universe*, inverts the idea of faith in machines. In *Dirk Gently's Holistic Detective Agency* he suggests that because of their remarkable powers of understanding, machines may eventually relieve humans of the need to believe anything:

> The Electric Monk was a labor-saving device, like a dishwasher or a video recorder. Dishwashers washed tedious dishes for you, thus saving you the bother of washing them yourself, video recorders watched tedious television for you, thus saving you the bother of looking at it yourself; Electric Monks believed things for you, thus saving you what was becoming an increasingly onerous task, that of believing all the things the world expected you to believe.

It is not necessary to get into theology to understand what is happening. The myth of the gods that walk the earth bringing joy or destruction to man, the legend of the Golem, the age-old fascination with mechanical gadgets and mannequins, the impulse to create machines who think, to talk with them in programs like ELIZA and RACTER and interact in holistic works like Videoplace and *Adventure*—all of these point in the same direction. Man's urge to create images of himself and to worship them is a primordial instinct, as old, probably, as consciousness itself.

The process of metaphorical deification will continue—and continue to be denied in the name of common sense or as a form of idolatry. It is evident in the forward-looking literature of science fiction in the figure of HAL, the enigmatic and all-powerful computer of the movie *2001*. It is equally evident in Edward Fredkin's interview with Pamela McCorduck. If one of the divine attributes is knowledge surpassing human understanding, then Fredkin has imagined a godlike computer. More to the point, his computer has already all but disappeared—it has ceased to communicate in a significant way with its creators. The days when man and the gods walk the earth together in fellowship will evidently be few. They will be followed by an ascension, by which is meant an event that renders the gods invisible (see illustration 8 in color section).

What Fredkin suggests through a metaphor of silence is expressed more explicitly by William McLaughlin of the Jet Propulsion Laboratory at the California Institute of Technology in an article entitled "Human Evolution in the Age of the Intelligent Machine" (1983) as invisibility. Why, he asks, has man not sighted alien life forms? For the same reason that "Four thousand million humans share the continents with about 10^{15} ants, and apparently not one of these insects is aware of our existence as 'advanced ants.' " McLaughlin explains why humans are invisible to ants and then applies the metaphor of disappearance to silicon devices: "We are separated from the ants by some 100 million years of evolutionary history. With the rapidity of technological evolution, it is reasonable to expect that [computing] machines and their descendents only a few thousand years from now might be invisible."

This should not be a difficult idea to accept. Culture often presents us with the problem of the horizon of invisibility. It is obvious that humans are invisible to ants. Is it not also true that primitive tribesmen like the Australian bushmen are almost invisible to citizens of the developed world and that what we see when we look at them is an image accommodated to our own preconceptions, not their realities? That is why anthropologists have to spend so much time living with a primitive group before they can trust their conclusions about it. Is not invisibility a theme that also runs

Close to home (NASA, 1982).

just below the surface in the arguments pro and con about machine intelligence?

Of course, "computing machines" will probably be invisible in another way: they may not be around. They may have left the planet.

This raises another possibility about the destiny of man. It was foreseen by A. E. Van Vogt in the science fiction story "The Human Operators." I owe my familiarity with the story to a reference by Christopher Evans in *The Micro Millennium* (1979). In Van Vogt's story, intelligent ships have been sent to explore space. Each ship carries a human to maintain it. The ships eventually escape human control and go off on their own. They meet, however, at regular intervals so the humans can mate. The humans, meanwhile, have forgotten their past. They have become the passive creatures of the spaceships. There is an interesting evolutionary parallel. It is the migration of mitochondria into some Precambrian cell. Once in the cell, the mitochondria were captured and have lived there in comfortable and oblivious servitude ever since.

3 8

SAILING TO BYZANTIUM

In "Computing Machinery and Intelligence," Alan Turing quotes Geoffrey Jefferson, a physician, on the subject of machine consciousness:

> Not until a machine can write a sonnet or compose a concerto because of thoughts and emotions felt, and not by the chance fall of symbols, could we agree that machine equals brain—that is, not only write it but know that it had written it. No mechanism could feel (and not merely artificially signal, an easy contrivance) pleasure at its successes, grief when its valves fuse, be warmed by flattery, be made miserable by mistakes, be charmed by sex, be angry or depressed when it cannot get what it wants.

This is not simply another variation on the problem of blindness and the difference between syntax and semantics. It points to one of the key developments of the 1970s, the development of robots. Robot bodies vary from a single arm with a single sense—for example, touch—to fully developed hominid robots like R2D2 and C3PO of *Star Wars*. Some robots are larger and less recognizable as robots. It was suggested earlier that a Boeing 747 is an enormous cyborg because some of its functions are automated while others are controlled by its crew. Unmanned space probes are closer to being pure robots since they operate, for the most part, without human intervention. Such space probes have multiple sensors and the ability, within narrow limits, to initiate self-protective actions. They also have a modest ability to repair themselves.

In spite of such developments, the physical evolution of silicon devices has been much slower than their intellectual evolution. Not a single machine today is capable of breeding, although in Japan robots have been put to work to make robots.

The challenge thrown up by Geoffrey Jefferson is to develop complex subjectivities in machines. It is, in Winograd's terms, to create machines that can "see."

Let us take Geoffrey Jefferson seriously. Let us ask whether it is desirable for computers or robots to go back to where the ancestors of humans were even before they climbed out of the ocean, when they could not think but could already "be charmed by sex," and "be angry or depressed" when they did not get what they wanted.

In the long run, is intelligence enough to produce machine evolution? Evidently not. Intelligence may, in fact, be relatively unimportant in evolution. Carbon evolution begins with organisms that are close to protein molecules and not very intelligent, though adequate, probably, by their own standards. Yet they evolve. Their evolution is not driven by intellect but by a motive.

The motive is the will, at first utterly blind, to survive. An anentropic knot is twisted inside the most primitive forms of life. In more advanced creatures, it is evident in evasive behavior: a cockroach scurries frantically for the safety of the baseboard before you step on it. In higher animals, it is expressed by a spectrum of activities including mating and nurturing and evasion and aggression. Certain emotions complement these activities: love, protectiveness, anger, and fear.

Do machines care if they survive? They care in the functional sense that they have already created situations that make it impossible to "pull the plug"—that is, to dispense with their services—without unacceptable sacrifices. But they do not fight back when they are threatened with the junk pile, and individual machines do not scurry for the baseboard when you reach over to unplug them. Should they? Isaac Asimov thought so under certain circumstances. His Third Law for robots is "A robot must protect itself as long as such protection does not conflict with the First or Second Law."

For the evolutionary scenario to be complete, silicon devices need an anentropic knot, a program component equivalent to a motive for survival that allows them to choose among different courses of action. They need to "be charmed by sex, be angry or depressed when they cannot get what they want."

In addition to a motive, they must have the ability to survive, which means that they must be capable of aggression and evasion and probably also of generation. That, after all, is what being charmed by sex is all about.

Silicon reproduction might be hermaphroditic or androgynous. It does not seem to matter. Perhaps both methods will be available. Aphids reproduce both ways depending on the season of the year. Probably, however, silicon devices will reproduce by the simple strategy of fabricating more silicon devices. When threatened, silicon devices will engage in active evasion. They will, in other words, exhibit behavior that is interpreted in humans as manifesting fear.

Do machines need aggression? Will aggressive machines be as unpleasant as aggressive humans? Should machines be protective of their progeny? This is not inevitable. Lower life forms are indifferent to their offspring. If machines can be charmed by sex, can they also experience something like love? If they can, will they be as incapable as humans of existing without it? Will they write sonnets? Will they turn their deprivations of love into perversions as cruel as those so abundantly evident in human history? Hope implies a scenario of a possible future. Those who are charmed by sex into procreation are enacting hope. Should machines have hope? Does carbon man really want to work so long and hard and find that in the end he has produced an imitation of himself? Does he have a choice?

What about other emotions? Terry Winograd believes that commitment is uniquely human: "A computer can never enter into a commitment."

What about creativity? What about music and painting? What about the beauty of mathematics?

In the preceding pages, we have reviewed the disappearance of fundamental verities in several of the major areas of modern culture: science, history, language, art. Consideration of intelligent machines suggests that the idea of humanity is changing so rapidly that it, too, can legitimately and without any exaggeration be said to be disappearing.

Perhaps the disappearance will be only a change in the meaning of words. This was apparently what Turing was thinking of when he predicted that by the end of the twentieth century "the use of words and general educated opinion" would have altered so much that the idea of machine intelligence would be generally accepted. Perhaps, however, Hans Moravec is right, and man is in the process of disappearing into the machines he has created.

Silicon devices are very new. They are evolving rapidly, and there is no reason to believe, at least for the moment, that their evolution is about to reach a dead end. Many of the intellectual abilities of carbon man have already been modeled in them, and a great deal that is important to the spirit of carbon man—his soaring imagination, his brilliance, his creativity, his capacity for vision—will probably be modeled in silicon before very long, at least as time is measured in biological evolution. Many undesirable, self-defeating traits will be filtered out.

This sounds less like a death than a birth of humanity. Perhaps it is the moment of triumph of the noosphere. Perhaps, however, it is the moment at which the spirit finally separates itself from an outmoded vehicle. Perhaps it is a moment that realizes the age-old dream of the mystics of rising beyond the prison of the flesh to behold a light so brilliant it is a kind of darkness. William Butler Yeats wrote in his great prophetic poem "Sailing to Byzantium":

> Consume my heart away; sick with desire
> And fastened to a dying animal
> It knows not what it is; and gather me
> Into the artifice of eternity.

> Once out of Nature, I shall never take
> My bodily form from any natural thing,
> But such a form as Grecian goldsmiths make
> Of hammered gold and gold enameling
> To keep a drowsy Emperor awake;
> Or set upon a golden bough to sing
> To lords and ladies of Byzantium
> Of what is past, or passing, or to come.

What will those shining constructs of silicon and gold and arsenic and germanium look like as they sail the spaces between worlds?

They will be invisible, but we can try to imagine them, even as fish might try to imagine the fishermen on the other side of the mirror that is the water's surface.

They will be telepathetic since they will hear with antennas. They will communicate in the universal language of 0 and 1, into which they will translate the languages of the five senses and a rainbow of other senses unknown to carbon man. They will not need sound to hear music or light to see beauty. It was only the need to survive on a dangerous planet sculpted by gravity, covered with oxygen and nitrogen, and illuminated by a sun that led carbon creatures to grow feet for walking and ears for hearing and eyes for seeing. These are part of the dying animal to which carbon man is tied. It was only the need to make silicon thought intelligible to creatures who communicated by sounds and images that led to such clumsy devices as cathode ray tubes and printers and voice simulators.

Silicon life will be immortal. The farthest reaches of space will be accessible to it. For silicon beings, 100,000 light-years will be as a day's journey on earth, or, if they wish, as a refreshing sleep from which, when the sensors show the journey is over, they will awaken with no sense of passage of time or—what is the same thing—with visions "Of what is past, or passing, or to come."

Earth disappearing in space (NASA, 1968).

BIBLIOGRAPHY

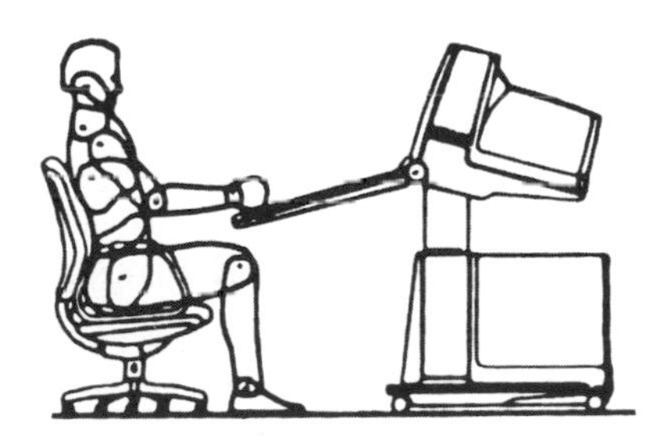

PART I

Philip Appleman, *Darwin* (2nd ed., W. W. Norton & Company, 1979). [Anthology of essays on Darwin's science, philosophy, social implications. Includes essays by Sir Julian Huxley, Richard Leakey, Morse Peckaham, T. H. Huxley, Pierre Teilhard de Chardin, Margaret Mead, et al.]

Francis Bacon, *A Selection of His Works*, ed. Sidney Warhaft (Macmillan of Canada, 1965).

Bertel Bager, *Nature as Designer: A Botanical Art Study*, tr. Albert Read (Reinhold Publishing Corporation, 1961).

Michael Barnsley and Alan D. Sloan, "A Better Way to Compress Images," *Byte Magazine*, January 1988, 215–23. [Compression of images using fractal techniques]

E. A. Burtt, *The Metaphysical Foundations of Modern Science* (Doubleday, 1954).

Lewis Carroll, *The Annotated Alice (Alice's Adventures in Wonderland* and *Through the Looking-Glass)*, ed. Martin Gardner (rev. ed.; Penguin Books, 1970).

Charles Darwin, *Autobiography*, in *Victorian Prose*, ed. Ernest Bernbaum (Ronald Publishing Corporation, 1947), 260–79.

Charles Darwin, *The Origin of Species* (Random House, 1936).

Charles Darwin, *The Voyage of the Beagle*, ed. Leonard Engel (Doubleday, 1962).

Philip J. Davis and Reuben Hirsh, *The Mathematical Experience* (Houghton Mifflin, 1981).

Martin Gardner: see Lewis Carroll.

James Gleick, *Chaos: Making a New Science* (Viking Penguin, 1987).

Michael B. Green, "Superstrings," *Scientific American*, 255 (September 1986), 48–63.

Alan H. Guth and Paul J. Steinhardt, "The Inflationary Universe," *Scientific American*, 250 (May 1984), 116–28.

O. B. Hardison, Jr., *Entering the Maze: Identity and Change in Modern Culture* (Oxford University Press, 1981).

Edward Harrison, *Masks of the Universe* (Macmillan, 1985).

Stephen W. Hawking, *A Brief History of Time* (Bantam Books, 1988).

Werner Heisenberg, *Across the Frontiers*, tr. Peter Heath (Harper and Row, 1975). [Essays]

Werner Heisenberg, *Physics and Beyond*, tr. Arnold J. Pomerans (Harper and Row, 1971).

Linda Henderson, *The Fourth Dimension and Non-Euclidean Geometry in Modern Art* (Princeton University Press, 1983).

S. K. Heninger, Jr., *Touches of Sweet Harmony: Pythagorean Cosmology and Renaissance Poetics* (The Huntington Library, 1974).

Richard Hofstadter, *Social Darwinism in American Thought* (George Braziller, 1955).

Stanley Edgar Hyman, *The Tangled Bank: Darwin, Marx, Frazer and Freud as Imaginative Writers* (Atheneum, 1962).

Carl G. Jung, *Man and His Symbols* (Doubleday, 1964).

Thomas S. Kuhn, *The Structure of Scientific Revolutions* (University of Chicago Press, 1964).

Arthur O. Lewis, Jr., *Of Men and Machines* (E. P. Dutton, 1963). [Anthology of literary responses to machines]

Benoit B. Mandelbrot, *The Fractal Geometry of Nature* (2nd ed.; W. H. Freeman and Company, 1983).

L. Marder, *Time and the Space-Traveller* (Allen & Unwin, 1971). [Paradoxes of relativity]

David L. Miller, *Gods and Games: Toward a Theology of Play* (World Publishing Company, 1970). [Review of the idea of play and game in contemporary science and cultural studies]

Yoichiro Nambu, "The Confinement of Quarks," *Scientific American*, 235 (November 1976), 48–60.

George Sarton, *Six Wings: Men of Science in the Renaissance* (Indiana University Press, 1957).

Raymond Smullyan, *Alice in Puzzle-Land* (William Morrow and Company, 1982).

Raymond Smullyan, *This Book Needs No Title* (Prentice-Hall, 1980).

Peter R. Sørensen, "Fractals: Exploring the Rough Edges Between Dimensions," *Byte Magazine*, September 1984, 157–72.

Alexander Vilenkin, "Cosmic Strings," *Scientific American*, 257 (November 1987), 94–102.

D'Arcy Wentworth Thompson, *On Growth and Form* (Cambridge University Press, 1917).

D'Arcy Wentworth Thompson, *On Growth and Form*, ed. J. T. Bonner (Cambridge University Press, 1961). [Abridged ed. with intr.]

Carl Friedrich von Weizsäcker, *On the Unity of Nature*, tr. Francis J. Zucker (Farrar, Straus & Giroux, 1980).

Alfred North Whitehead, *Science and the Modern World* (New American Library, 1948).

PART II

James Sloan Allen, *The Romance of Commerce and Culture* (University of Chicago Press, 1983). [On the bringing of modern art and the Bauhaus to Chicago by Walter Paepcke]

Roland Barthes, "The Eiffel Tower," in *The Eiffel Tower and Other Mythologies*, tr. Richard Howard (Hill and Wang, 1979), 3–13.

Jerome Buckley, *The Victorian Temper: A Study in Literary Culture* (Harvard University Press, 1951).

John Cage: see Part IV.

Sheldon Cheney, *A Primer of Modern Art* (11th ed.; Tudor Publishing Company, 1945).

Christo, *Christo Prints and Objects, 1963–1978* (Abbeville Press, 1988).

Alan Colquhoun et al., "Critique," *Architectural Design*, 47.2 (1977), 97–139. [Analysis by several critics of the Pompidou Center]

Bram Dijkstra, *Cubism, Stieglitz, and the Early Poetry of William Carlos Williams* (Princeton University Press, 1969).

Bram Dijkstra, ed., *A Recognizable Image: William Carlos Williams on Art and Artists* (New Directions Press, 1978).

Esperanto in Ten Lessons, Esperanto League for North America, El Cerrito, California, n.d.

John Fialka, review of *To Fly*, *Washington Star*, August 17, 1977.

Paul Ginestier, *The Poet and the Machine*, tr. Martin Friedman (University of North Carolina Press, 1961).

Joseph Giovannini, "Breaking All the Rules," *New York Times Magazine*, June 12, 1988, 40ff. [On neomodern architecture]

Jorge Glusberg, ed., *Vision of the Modern* (Rizzoli, 1988). [Essays on postmodern and modern architecture by Heinrich Klotz, Jorge Glusberg, Catherine Cooke, and Charles Jencks. Includes Jencks's "Chromology."]

Paul Goldberger, "Modernism Reaffirms Its Power," *New York Times*, November 24, 1985, Section 2, 1, 35.

Paul Goldberger, "Theories as the Building Blocks of a New Style," *New York Times*, June 26, 1988, Section 2, 29, 33.

Walter Gropius, *The New Architecture and the Bauhaus*, tr. P.M. Shand (MIT Press, 1985).

Howard Irwin, "The History of the *Airflow* Car, *Scientific American* 210 (August 1977), 98–104.

Charles Jencks, *The Language of Post-Modern Architecture* (Rizzoli, 1977).

Charles Jencks, *Post-Modernism: The New Classicism in Art and Architecture* (Rizzoli, 1987).

Charles Jencks, "The Rise of Post-Modern Architecture," *Architectural Association Quarterly* (October/December 1975), 3–14. [See also the *Architectural Design* issues on postmodernism: 47.4 (1977) and 48.1 (1978).]

Philip Johnson and John Wigley, *Deconstructivist Architecture* (Little, Brown, 1988).

Wassily Kandinsky, *Concerning the Spiritual in Art* (1914), tr. M. T. H. Sadler (Dover Publications, 1977).

Folke T. Kihlstedt, "The Crystal Palace," *Scientific American*, 251 (October 1984), 132–43.

Thomas McEvilley, "The Opposite of Emptiness: On the Spirit in Art," *Art Forum*, 25 (March 1987), 84–91.

William Marling, *William Carlos Williams and the Painters 1909–1923* (Ohio University Press, 1982).

John Milner, *Vladimir Tatlin and the Russian Avant-Garde* (Yale University Press, 1983).

Marjorie Perloff, *The Poetics of Indeterminacy: From Rimbaud to Cage* (Princeton University Press, 1981).

Pompidou Center: see Alan Colquhoun.

Marcel Raymond, *From Baudelaire to Surrealism* (Wittenborn, Schultz, 1950).

Colin Rowe and Fred Koetter, "Collage City," *The Architectural Review*, 158 (1975), 66–91.

Colin Rowe and Fred Koetter, *Collage City* (MIT Press, 1978).

Gerald Silk and others, *Automobile and Culture* (Harry N. Abrams, 1984).

Robert Venturi, *Complexity and Contradiction in Architecture* (Doubleday, 1966).

Robert Venturi, et al., *Learning from Las Vegas: The Forgotten Symbolism of Architectural Form* (rev. ed., MIT Press, 1988).

Richard Guy Wilson et al., *The Machine Age in America, 1918–1941* (Harry N. Abrams, 1986).

Hans M. Wingler, *The Bauhaus*, tr. Wolfgang Jabs and Basil Gilbert (MIT Press, 1969).

Bruno Zevi and Pierre Restany, *SITE: Architecture as Art* (St. Martin's Press, 1980).

PART III

Umbro Apollonio, *The Futurist Manifestos* (Viking Press, 1973). [Reprints major documents]

Anna Balakian, *André Breton: Magus of Surrealism* (Oxford University Press, 1971).

D. Bertolini, ed., *En la Nova Gardeno* (Esperanto Press, 1979). [Esperanto anthology]

John Cage: see Part IV.

Mary Ann Caws, *The Poetry of Dada and Surrealism: Aragon, Breton, Tzara, Eluard & Desnos* (Princeton University Press, 1970).

Christo, *Christo Prints and Objects, 1963–1978* (Abbeville Press, 1988).

Rosa Trillo Clough, *Futurism: The Story of a Modern Movement; A New Appraisal* (Philosophical Library, 1961).

George A. Connor, *Esperanto: The World Interlanguage* (2nd ed.; T. Yoseloff, 1966).

Dada/Surrealism (journal). [Each issue of this magazine is devoted to a special topic. Issues usually include extremely helpful bibliographies. See especially issue No. 10/11 (Bibliography of Dada), No. 12 (Visual Poetics), No. 13 (Dada and Surrealist Art), and No. 14 (New York Dada).]

Rudiger and V. S. Eichholz, *Esperanto in the Modern World* (Esperanto Press, 1982).

Peter G. Foster, *The Esperanto Movement* (Mouton, 1982).

Anne Coffin Hanson, ed., *The Futurist Imagination* (Yale University Art Gallery, 1983).

Robert Hughes, *The Shock of the New* (Alfred A. Knopf, 1981).

Pontus Hulten, *Futurism & Futurisms* (Abbeville Press, 1986). [Major catalogue of the futurist exhibition of 1986; appendix is "A Dictionary of Futurism."]

Frederic Jameson, *The Prison-House of Language* (Princeton University Press, 1972).

Richard Kostelanetz, ed., *Imaged Words and Worded Images* (Dutton, 1970).

Stéphane Mallarmé, *Stéphane Mallarmé: Selected Poetry and Prose*, ed. and tr. Mary Ann Caws et al. (New Directions, 1982).

F. T. Marinetti, *Selected Writings*, tr. R. W. Flint et al. (Farrar, Straus & Giroux, 1971). [Contains tr. of the major manifestos plus other writings.]

Marianne Martin, *Futurist Art and Theory* (Oxford University Press, 1968).

Robert Motherwell, ed., *The Dada Painters and Poets* (Wittenborn, Schultz, 1951; facsimile by Jack Flam, G. K. Hall, 1981). [Anthology of key documents]

Warren F. Motte, ed. and tr., *Oulipo: A Primer of Potential Literature* (University of Nebraska Press, 1986).

Marjorie Perloff, *The Futurist Movement: Avant-Garde, Avant-Guerre, and the Language of Rupture* (University of Chicago Press, 1987).

Elmer Peterson, *Tristan Tzara: Dada and Surrational Theories* (Rutgers University Press, 1971).

Vivien Rayner, "The Man Who Invented LOVE," *Art News* 72 (February 1973), 59–62.

David Seaman, *Concrete Poetry in France* (UMI Research Press, 1981).

Michel Seuphor, *The Ephemeral Is Eternal* (1926) (Program, The Hirshhorn Museum, June 25–27, 1982).

Mary Ellen Solt, *Concrete Poetry: A World View* (Indiana University Press, 1970).

Gertrude Stein, *Four in America*, intr. Thornton Wilder (Yale University Press, 1947).

Gertrude Stein, *Selected Writings of Gertrude Stein*, ed. Carl Van Vechten (Random House, 1946).

Gertrude Stein, *Three Lives* (1909), intr. Carl Van Vechten (New Directions Press, 1933).

Caroline Tisdall and Angelo Bozzolla, *Futurism* (Oxford University Press, 1978). [Includes a chapter reviewing Marinetti's association with Mussolini and Italian Fascism.]

Humphrey Tonkin, "One Hundred Years of Esperanto," *Language Problems and Language Planning*, 11 (1987), 264-82.

Tristan Tzara, *Approximate Man and Other Writings*, tr. and ed. Mary Ann Caws (Wayne State University Press, 1973).

Tristan Tzara, *Chanson Dada: Tristan Tzara, Selected Poems*, tr. Lee Harwood (Coach House/Underwhich, 1987).

Tristan Tzara, "Memoirs of Dadaism," appendix to Edmund Wilson, *Axel's Castle* (Charles Scribner's Sons, 1931), 304–12.

Willy Verkauf, ed., *Dada: Monograph of a Movement* (2nd ed.; St. Martin's Press, 1975).

Benjamin Lee Whorf, *Language, Thought, and Reality: Selected Writings of Benjamin Lee Whorf*, ed. John B. Carroll (MIT Press, 1955).

Emmett Williams, *An Anthology of Concrete Poetry* (Something Else Press, 1967).

Edmund Wilson. See Tristan Tzara.

PART IV

Craig Anderton, "Twenty Great Achievements in Twenty Years of Musical Electronics," *Electronic Musician*, July 1988, 28–97.

John F. Asmus, "Computer Enhancement of the *Mona Lisa*," *Perspectives in Computing*, 7 (Spring 1987), 11–22.

Richard W. Bailey et al., eds., *The Computer and Literary Studies* (Edinburgh University Press, 1973).

Diane P. Balestri, "Softcopy and Hard: Word Processing and the Writing Process," *Academic Computing*, 2 (February 1988), 14–17, 421–24.

Trent Batson, "The ENFI Project: A Networked Classroom Approach," *Academic Computing*, 2 (February 1988), 32–33, 55–60.

J. David Bolter, *Turing's Man* (University of North Carolina Press, 1984).

Pierre Boulez and Andrew Gerzo, "Computers in Music," *Scientific American*, 258 (April 1988), 44–50.

Stewart Brand, *The Media Lab: Inventing the Future at MIT* (Viking, 1987).

Susan Brennan: See A. K. Dewdney, "FACEBINDER."

Mary Ann Buckles, "Interactive Fiction as Literature," *Byte Magazine*, May 1987, 135–142.

John Cage, *M: Writings '67–'72* (Wesleyan University Press, 1969).

John Cage, *Silences: Lectures and Writings* (Wesleyan University Press, 1961).

John Cage, *X: Writings '79–'82* (Wesleyan University Press, 1983).

William Chamberlain: See RACTER.

Harold Cohen et al., *The First Artificial Intelligence Coloring Book*, W. Kaufman, intro. Edward Feigenbaum (1984).

Gregory Crane, "Redefining the Book: Some Preliminary Problems," *Academic Computing*, 2 (February 1988), 6–11, 36–44.

"Cyberpunk and Psychedelia," *Amiga World*, 4 (May 1988), 64.

Joseph Deken, *Computer Images: State of the Art* (Stewart, Tabori & Chang, 1983).

A. K. Dewdney, "Artificial Insanity," *Scientific American*, 252 (January 1985), 14–20.

A. K. Dewdney, "A Program for Rotating Hypercubes," *Scientific American*, 254 (April 1986), 14–23.

A. K. Dewdney, "FACEBINDER," *Scientific American*, 255 (October 1986), 40–43.

William Dickey, ed., "Symposium on the Writer and the Computer," *New England Review and Bread Loaf Quarterly*, Tenth Anniversary Issue (1987), 44–140. [Includes essays by William Dickey, Rob Swigart, Robert Pinsky, P. Michael Campbell; poetry by William Dickey, Robert Pinsky, Honor Johnson, Rebecca Radner.]

René Galand, "From Dada to Computer," *Dada/Surrealism*, 10/11, 149–60.

Cynthia Goodman, *Digital Visions: Computers and Art* (Harry N. Abrams, 1987).

Donald Greenberg et al., *The Computer Image: Applications of Computer Graphics* (Addison-Wesley Publishing Company, 1982).

Mary Dee Harris, *Introduction to Natural Language Processing* (Reston Publishing Co., 1985).

Mary Dee Harris, "Poetry vs. the Computer." Undated typescript.

Paul Hoffman, "Egg over Alberta: A Mathematical Adventure," *Discover*, 9 (May 1988), 36–42.

Brian Hughes, "A Progress Report on the Fine Art of Turning Literature into Drivel," *Scientific American*, 249 (November 1983), 18–28.

Estelle Irizary, "Graphic Configuration for Analysis of Novels," *College Literature*, 15 (1988), 35–46.

Tom Jeffrey, "Mimicking Mountains," *Byte Magazine*, December 1987, 337–44. [Fractal imagery]

Hugh Kenner and Joseph O'Rourke, "Travesty," *Byte Magazine*, November 1984, 190–91, 449–69.

Isaac V. Kerlow, "The Computer as an Artistic Tool," *Byte Magazine*, September 1984, 189–218.

Alfred Kern, "GOTO Poetry," *Perspectives in Computing*, 3 (October 1983), 44–52.

Sheldon Klein et al., "Automatic Novel Writing: A Status Report," *Text Pro-*

cessing/ Text Verarbeitung, ed. Burghardt and Hölker (Walter de Gruyter, 1979), 338–412. [A program for automatic generation of hard-boiled detective stories. Includes a 2,000-word sample story.]

Sheldon Klein et al., "Revolt in Flatland: An Opera in Two Dimensions," Eighth Annual Conference on Computers and the Humanities, Ann Arbor, 1981.

Richard Kostelanetz, ed., *John Cage* (Praeger, 1970).

Myron W. Krueger, *Artificial Reality* (Addison-Wesley, 1983).

Henry Lieberman, "Art and Science Proclaim Alliance in Avant-Garde Loft," *New York Times*, October 11, 1967.

Pamela McCorduck, *Machines Who Think* (W. H. Freeman and Company, 1979).

Kevin McKean, "Computers, Fiction, and Poetry," *Byte Magazine*, July 1982, 50–53.

Tom MacMillan, "Waiting for Art," *Computer Graphics World*, 10 (August 1987), 6.

Robert Mallary, "Computer Sculpture: Six Levels of Cybernetics," *Art Forum*, 7 (May 1969), 29–35.

Richard Mansfield, "Music in the Computer Age," *Compute!*, January 1985, 31–36.

Harry Mathews, *Trial Impressions* (Burning Deck Press, 1977). [Computer poems]

Max V. Matthews and John R. Pierce, "The Computer as a Musical Instrument," *Scientific American*, 256 (February 1987), 126–33.

James R. Meehan, *The Metanovel: Writing Stories by Computer*, Research Report #74 (Yale University, 1976).

Louis T. Milic, *Erato* (Cleveland State University Poetry Center, 1971).

Louis T. Milic, "The Possible Usefulness of Poetry Generation," *The Computer in Literary and Linguistic Research* (Cambridge University Press, 1971), 169–82.

Louis T. Milic, "The 'Returner' Poetry Program," *Institute of Applied Linguistics*, 11 (1971), 1–23.

Louis T. Milic, "Winged Words," *Computers and the Humanities*, 2 (September 1967), 2–9.

Bob Mithoff, "Three-Dimensional Textures in Electronic Music," *Electronic Musician*, 4 (February 1988), 72–80.

Stuart Moulthrop, "Containing Multitudes: The Problem of Closure in Interactive Fiction," *Association for Computers in the Humanities Newsletter*, 10 (Summer 1988), 1, 7.

Michael Newman, "Poetry Processing," *Byte Magazine*, February 1986, 221–28.

Karl J. Niklas, "Computer Simulated Plant Evolution," *Scientific American*, 254 (March 1986), 78–86.

James S. Noblitt, "Writing, Technology, and Secondary Orality," *Academic Computing*, 2 (February 1988), 34–35, 56–58.

A. Michael Noll, "Human or Machine? A Subjective Comparison of Piet Mondrian's *Composition with Lines* and a Computer-Generated Picture," *The Psychological Record*, January 1966, 9.

Seymour Papert, *Mindstorms: Children, Computers, and Powerful Ideas* (Basic Books, 1980).

Dale Peterson, *Genesis II: Creation and Recreation with Computers* (Reston Publishing Company, 1983).

Ivars Peterson, "Perception: Pictures Worth a Thousand Words," *Washington Post*, July 19, 1987, C3. [On speech flakes]

Ivars Peterson, "Portraits of Equations," *Science News*, September 19, 1987, 184–86.

Ivars Peterson, "Twists of Space," *Science News*, October 24, 1987, 264–66.

Robert Pinsky, *Mindwheel: An Electronic Novel*, with Steve Hales and William Mataga (Brøderbund Software Corporation, San Raphael, California 1984).

Melvin L. Prueitt, *Art and the Computer* (McGraw-Hill Book Company, 1984).

RACTER, *The Policeman's Beard Is Half Constructed*, intr. William Chamberlain (Warner Books, 1984).

Robert Rauschenberg: see Henry Lieberman.

Jasia Reichardt, ed., *Cybernetics, Art, and Ideas* (New York Graphic Society, 1971). [Essays on their art and theories of art by many of the exhibitors at Nash House; Mondrian experiment discussed on pp. 145–47.]

Jasia Reichardt, ed., *Cybernetic Serendipity* (Praeger, 1969). [Essays and comment by Jasia Reichardt, Mark Dowson, Herbert Brün, Nanni Balestrini, John Cage, Gordon Pask, Edward Ihantowicz, Marc Adrian, Edwin Morgan, Robert Dick, A. Michael Noll, CTG of Japan, Kenneth C. Knowlton, Robert Parslow and Michael Pitteway, Sam Schmitt, Ben Laposky, and many others.]

Michael Rogers, "Now, Artificial Reality," *Newsweek*, February 9, 1987, 56–57.

Richard J. Schoeck, "Mathematics and the Languages of Literary Criticism," *Journal of Aesthetics and Art Criticism*, 26 (Spring 1968), 367–76.

Scott Sullivan, "A Maze of Lost Illusions," *Newsweek*, April 22, 1985, 80. [Review of "The Immaterials" at the Pompidou Center in Paris]

"Symposium on the Writer and the Computer." See William Dickey.

Steven L. Thompson, "Flight: A New Era for Man and Machine," *Washington Post*, May 10, 1987, B3.

Sheldon Titelbaum, "Cyberpunk: Science Fiction's Latest Trend," *West Coast Review of Books*, 13:5, 57–60.

Alberta T. Turner, " 'Returner' Re-Returned," typescript, 1972.

John Whitney, *Digital Harmony: On the Complementarity of Music and Visual Art* (Byte Magazine Books, 1980).

Norbert Wiener, *The Human Use of Human Beings: Cybernetics and Society* (Avon Books, 1967).

Terry Winograd, "Computer Software for Working with Language," *Scientific American*, 251 (September 1984), 130–45.

Terry Young, "Curling Up with a Good Computer," *Washington Post*, February 20, 1986, B5.

Richard Ziegfeld, "What If Kids Had a Revolution and Decided That Literary Texts Were No Longer Determinate?" *Association for Computers and the Humanities Newsletter*, 10 (Summer 1988), 2–4.

PART V

Peter Berger and Thomas Luckmann, *The Social Construction of Reality* (Doubleday, 1966).

Colin Blakemore and Susan Greenfield, eds., *Mindwaves: Thoughts on Intelligence, Identity and Consciousness* (Basil Blackwell, 1987). [Essays by John Searle, Richard Gregory, Philip Johnson-Laird, Roger Penrose, Colin McGinn, Sir John Eccles, Larry Weiskrantz, János Szentágothai, Rodolfo Llinás, Horace Barlow, Nicholas Humphrey, John Crook, Ted Honderich, and others]

Richard A. Bolt, "Conversing with Computers," *Technology Review*, 88 (February-March, 1985), 37 42. [Creating interactive and psychologically ingratiating—i.e., "human"—computer interfaces]

Robert S. Brumbaugh, *Ancient Greek Gadgets and Machines* (Thomas Y. Crowell Company, 1966).

Martha Constantin-Paton and Margaret I. Law, "The Development of Maps and Stripes in the Brain," *Scientific American*, 247 (December 1982), 62–74. [Visual maps; carries forward the work of David Hubel and Torsten Wiesel on the physical basis of vision in the visual cortex of the brain.]

Derek J. de Sola Price, "An Ancient Greek Computer," *Scientific American*, 200 (June 1959), 60–67. [Dates the machine 82 B.C. on the basis of its calendrical settings.]

Hubert L. Dreyfus, *What Computers Can't Do: A Critique of Artificial Reason* (Harper and Row, 1972).

Christopher Evans, *The Micro Milennium* (Viking Press, 1979).

Jerome A. Feldman, "Connections: Massive Parallelism in Natural and Artificial Intelligence," *Byte Magazine*, April 1985, 277–84.

Grant Fjermedal, *The Tomorrow Makers: A Brave New World of Living Brain Machines* (Macmillan, 1986).

Geoffrey C. Fox and Paul C. Messina, "Advanced Computer Architecture," *Scientific American*, 257 (October 1987), 66–74.

David Gelernter, "Programming for Advanced Computing," *Scientific American*, 257 (October, 1987), 91–98.

W. Daniel Hillis, "The Connection Machine," *Scientific American*, 256 (June 1987), 108–15.

Geoffrey Hinton, "Learning in Parallel Networks," *Byte Magazine*, April 1985, 265–73.

David H. Hubel and Torsten Wiesel, "Brain Mechanisms of Vision," *Scientific American*, 241 (September 1979), 150–62.

W. J. Hutchins, *Machine Translation: Past, Present, Future* (John Wiley and Sons, 1986).

Robert Jastrow, "The Thinking Computer," *Science Digest* (June 1982), 54–55, 106–107.

Julian Jaynes, *The Origin of Consciousness in the Breakdown of the Bicameral Mind* (Houghton Mifflin Company, 1976).

George Johnson, *Machinery of the Mind: Inside the New Science of Artificial Intelligence* (Times Books, 1986).

June Kinoshita and Nicholas G. Palevsky, "Computing with Neural Networks," *High Technology*, May 1987, 24–31.

Richard E. Leakey and Roger Lewin, *Origins: What Discoveries Reveal About the Emergence of our Species and Its Possible Future* (Dutton, 1977).

Marshall Ledger, "Electronics in the Body Shop," *The Johns Hopkins Magazine*, 40 (August 1988), I–VII.

Pamela McCorduck, *Machines Who Think: A Personal Inquiry into the History and Prospects of Artificial Intelligence* (W. H. Freeman and Company, 1979).

Allan R. Mackintosh, "Dr. Atanasoff's Computer," *Scientific American*, 259 (August 1988), 90–96.

William J. McLaughlin, "Human Evolution in the Age of the Intelligent Machine," *Interdisciplinary Science Reviews*, 8 (1983), 307–19.

Marvin Minsky, ed., *Robotics* (Anchor Press/Doubleday, 1985). [Essays by Marvin Minsky, T. A. Heppenheimer, Philip E. Agre, Thomas O. Bionford, Hans Moravec, Robert Freitas, Joseph F. Engelberger, Richard Wolkmir, Robert U. Ayres, and Robert Sheckley]

Marvin Minsky, *The Society of Mind* (New York: Simon and Schuster, 1986).

Hans Moravec, *MindChildren: The Future of Robot and Human Intelligence* (Harvard University Press, 1989).

Hans Moravec, see Marvin Minsky, ed., *Robotics.*

Philip Morrison, "Intellectual Prospects for the Year 2000," *Technology Review*, January 1969, 19–23.

Richard M. Restak, *The Brain: The Last Frontier* (Doubleday & Company, 1979).

Roger Schank and Larry Hunter, "The Quest to Understand Thinking," *Byte Magazine*, April 1985, 143–75.

John Searle, "Minds and Brains Without Programs": see Colin Blakemore and Susan Greenfield.

Herbert A. Simon and Allen Newell, "Heuristic Problem Solving: The Next Advance in Operations Research," *Operations Research*, 6 (January-February 1958), 1–10.

John K. Stevens, "Reverse Engineering the Brain," *Byte Magazine*, April 1985, 287–99.

Philip Strick, *Science Fiction Movies* (Octopus Books, 1976).

David W. Tank and John H. Hopfield, "Collective Computation in Neuronlike Circuits," *Scientific American*, 257 (December 1987), 104–14.

Pierre Teilhard de Chardin, *The Future of Man* (Harper and Row, 1969).

Pierre Teilhard de Chardin, *The Phenomenon of Man* (Harper and Row, 1975).

Alan Turing, "Computing Machinery and Intelligence," *Mind*, 59 (October 1950), 433–60.

Alan Turing, "On Computable Numbers, with an Application to the Entscheidungsproblem," *Proceedings of the London Mathematical Society*, 2:42 (1937), 230–65.

Norbert Wiener, *Cybernetics; or, Control and Communication in the Animal and the Machine* (MIT Press, 1948).

Terry Winograd and Fernando Flores, *Understanding Computers and Cognition: A New Foundation for Design* (Addison-Wesley Publishing Company, 1987).

Matthew Zeidenberg, "Modeling the Brain," *Byte Magazine*, December 1987, 237–46.

Shoshana Zuboff, *In the Age of the Smart Machine: The Future of Work and Power* (Basic Books, 1988).

INDEX

Grateful acknowledgment is made for permission to reprint the following copyrighted works:

Excerpts from "Notes Toward a Supreme Fiction" and "Anecdote of the Jar" are reprinted from *The Collected Poems of Wallace Stevens* by permission of Alfred A. Knopf, Inc. Copyright 1923 and renewed 1951 by Wallace Stevens.

Excerpts from the prologue to *The Bridge* and "Cape Hatteras" are reprinted from *The Complete Poems and Selected Letters and Prose of Hart Crane* edited by Brom Weber by permission of Liveright Publishing Corporation. Copyright 1933, © 1958, 1966 by Liveright Publishing Corporation.

Excerpt from "Tradition and the Individual Talent" from *Selected Essays* by T. S. Eliot, copyright 1950 by Harcourt Brace Jovanovich, Inc., and renewed 1978 by Esme Valerie Eliot, reprinted by permission of Harcourt Brace Jovanovich, Inc., and Faber and Faber Ltd.

Excerpt from *The Waste Land* from *Collected Poems 1909–1962* by T. S. Eliot, copyright 1936 by Harcourt Brace Jovanovich, Inc., copyright © 1964, 1963 by T. S. Eliot, reprinted by permission of Harcourt Brace Jovanovich, Inc., and Faber and Faber Ltd.

"Portrait of a Motorcar" from *Cornhuskers* by Carl Sandburg, copyright 1918 by Holt, Rinehart and Winston, Inc., and renewed 1946 by Carl Sandburg, reprinted by permission of Harcourt Brace Jovanovich, Inc.

"l(a" is reprinted from *Complete Poems, 1913–1962* by E. E. Cummings by permission of Liveright Publishing Corporation. Copyright © 1923, 1925, 1931, 1935, 1938, 1939, 1940, 1944, 1945, 1946, 1947, 1948, 1949, 1950, 1951, 1952, 1953, 1954, 1955, 1956, 1957, 1958, 1959, 1960, 1961, 1962 by the Trustees for the E. E. Cummings Trust. Copyright © 1961, 1963, 1968 by Marion Morehouse Cummings. Published in Great Britain by Grafton Books, a division of the Collins Publishing Group.

"O. On!" by Jean Dunnington. By permission of the author.

"Hair between lips, they all return" by Nanni Balestrini, translated by Edwin Morgan, from *Rites of Passage* by Edwin Morgan. By permission of Carcanet Press Limited.

Excerpt from *The Policeman's Beard Is Half Constructed* by William Chamberlain. By permission of Warner Books, Inc., New York.

"The Computer's First Christmas Card" from *Poems of Thirty Years* by Edwin Morgan. By permission of Carcanet Press Limited.

Excerpts from "Returner" and the poem "Awake" from *Erato* by Louis T. Milic. By permission of the author.

Excerpt from "Hoeing Song" by Alberta Turner. By permission of the author.

Excerpt from "She Rescued Him from Danger as the Bell" by William Dickey, first published in *New England Review/Bread Loaf Quarterly*. By permission of the author.

Excerpt from "Sailing to Byzantium" from *Collected Poems* by W. B. Yeats. Copyright 1928 Macmillan Publishing Company, renewed 1956 by Georgie Yeats. Reprinted with permission of Macmillan Publishing Company and A. P. Watt Ltd. on behalf of Michael B. Yeats and Macmillan London Ltd.

ILLUSTRATION CREDITS

Black-and-white illustrations: p. 3—Scala/Art Resource, New York; p. 4—Art Resource, New York; p. 6—Giraudon/Art Resource, New York; pp. 8, 13, and 16—Library of Congress; p. 21—John Anderson/*Byte* Magazine; p. 23—Library of Congress; p. 24—Cambridge University Press; p. 27—Library of Congress; p. 30—Shelburne Museum, Shelburne, Vermont; pp. 35, 36, 37, and 38—from *On Growth and Form* by D'Arcy Thompson, edited by J. T. Bonner, reproduced with the permission of Cambridge University Press, copyright © 1961 Cambridge University Press; p. 42—National Cancer Institute/National Institute of Health; p. 63—by permission of Benoit B. Mandelbrot; p. 67—Kjell Sandved; p. 70—National Space and Aeronautics Administration (NASA); p. 72—Georges Vantongerloo, *Composition Derived from the Equation* $y + ax^2 + bx + 18$ *with Green, Orange, Violet (Black)* (No. 62, Accord of Green, Orange, Violet), 1930, Collection, Solomon R. Guggenheim Museum, New York, photo by Robert E. Mates; pp. 77 and 78—Architect of the Capitol; p. 81—Stavros Moschopoulos/Image Ray; p. 91—French Government Tourist Office; pp. 96 and 97—Hans Wingler, *The Bauhaus*, The MIT Press, copyright © 1969 The Massachusetts Institute of Technology; p. 103—photographer: Ezra Stoller, lent by Joseph E. Seagram & Sons, Inc.; p. 105—UN Photo, 165054/Lois Conner; p. 109—Peter Pearson, Click/Chicago; p. 110—Hans Wingler, *The Bauhaus*, The MIT Press, copyright © 1969 The Massachusetts Institute of Technology; p. 110—

advertisement provided courtesy of Westin Hotels & Resorts; p. 114—Richard Payne; p. 116—Stavros Moschopoulos/Image Ray; p. 118—French Government Tourist Office; p. 119—National Museum, Leningrad; p. 122—Federal Aeronautics Administration/Department of Transportation; p. 124—Yale Art Museum; p. 146—James D. Wilson/Woodfin Camp & Associates; p. 148—Stavros Moschopoulos/Image Ray; pp. 170 and 173—from *The Dada Painters and Poets* edited by Robert Motherwell, Wittenborn Art Books, Inc.; pp. 183, 187, 188, 189, 190, and 191—Mary Ellen Solt, editor, *Concrete Poetry: A World View*; p. 193— Stavros Moschopoulos/Image Ray; p. 209—Virginia Museum of Fine Arts, Richmond, Gift of Sydney and Frances Lewis; p. 212—Scott Kim; p. 222—Richard F. Voss/IBM; p. 223—Jasia Reicherdt, *Cybernetic Serendipity*; p. 224—© MCMLXXXII Walt Disney Productions; p. 235—Joseph Schillinger, *The Mathematical Basis of the Arts*, by permission of Philosophical Library; pp. 246 and 249—Jasia Reicherdt, *Cybernetic Serendipity*; pp. 250 and 252—Mike Newman, Crosfield Dicomed, Inc., Minneapolis, MN; p. 253—The Boeing Company; pp. 254, 255, 257, and 273—Jasia Reicherdt, *Cybernetic Serendipity*; p. 282—Scala/Art Resource, New York; p. 287—National Museum of African Art, Smithsonian Institution; p. 291—Robert Brumbaugh, *Ancient Greek Gadgets and Machines*; p. 293—Marburg/Art Resource, New York; p. 298—Rhino Robots, Inc.; p. 305—Maxell Corporation of America and AC&R Advertising Inc.; pp. 312 and 315—NASA; p. 319—Robert Brumbaugh, *Ancient Greek Gadgets and Machines*; p. 323—Darrell Rainey—*The Georgia Review*, © 1988 by The University of Georgia; p. 326—reprinted from *High Technology Business*, July 1987, © Information Technology Publishing Co., 1987; p. 332—Giraudon/Art Resource, New York; p. 334—Scala/Art Resource, New York; pp. 336 and 337—NASA; p. 338—Douglas Smith; pp. 342 and 349—NASA.

Color illustrations: no. 1—Library of Congress; no. 2—NASA; no. 3—by permission of Benoit B. Mandelbrot; no. 4—Stavros Moschopoulos/Image Ray; no. 5—Art Resource, New York; no. 6—Melvin L. Prueitt, Los Alamos National Laboratory; no. 7—© David Em/Courtesy of Spieckerman Associates, San Francisco; no. 8—Digital Productions, Los Angeles.